The Book of Time

Consultant editor Colin Wilson

Edited by John Grant

WESTBRIDGE BOOKS

A Division of David & Charles

British Library Cataloguing in Publication Data

The book of time.
  1. Time
  I. Grant, John
  529   QB209

ISBN 0–7153–7764–7

Printed in the United States of America

Published in the United States of America
by David & Charles Inc
North Pomfret Vermont 05053 USA

Contents

1 THE HISTORY OF TIME
Roy Porter   5

2 THE MOVING EARTH IN SPACE
Richard Knox   45

3 FROM SUNDIAL TO ATOMIC CLOCK
Chris Morgan   84

4 BODYTIME
E W J Phipps   129

5 MUTABLE TIME
Iain Nicolson   157

6 MEASURING TIME PAST
Brian John   236

7 TIME IN DISARRAY
Colin Wilson   285

   ABOUT THE AUTHORS   315

   ACKNOWLEDGEMENTS   317

   INDEX   318

1 The History of Time

"Creatures of an inferiour nature are possest with the *present*," declared the poet, John Donne: "*Man* is a *future Creature*." Donne had in mind the Christian belief in Man's heavenly destiny, but he had seized on a basic distinction between Man and practically all other animals: Man has ingrained senses of memory and anticipation. He orders his life within a grid of past, present and future.

This time sense dates back to primeval cultures. Neanderthal man (about 50,000BC) was already burying his dead (no other animal does that), which probably means that he thought the departed had some continued existence. The necessities of a life to come—food, tools, and weapons—were placed beside the body in burials from the Upper Palaeolithic period (about 35,000BC). From early times men have practised cults of ancestor worship. In myths and ritual the collective memory has re-enacted natural disasters, migrations, wars, etc. The megaliths of Stonehenge (begun about 1900BC) may well have comprised a sophisticated time computer. The earliest surviving inscription of events, the Palermo Stone from Egypt (*c.*2500BC), witnesses regular recording of the reigns of pharoahs and the floods of the Nile.

There is obviously, then, something universal in the need to escape the prison of the present. To secure food men must learn from experience; to maintain social cohesion they must validate the present by reference to tradition. They must also be able to anticipate and control the future. Basic tasks like maintaining fire require great forethought. The patient process of grinding tools means planning ahead. Religious sacrifice and propitiation pre-empt the future by trying to manipulate it. Furthermore, daily experience of birth, life and death—especially the mystery and terror of mortality—surely prompts belief in a magical world filled with spirits and the souls of the departed.

Yet notions of time and history are not innate. Very young infants live only in the present: the past is forgotten, the future

Stonehenge, the world's most famous example of a megalithic monument, thought to have been erected as an astronomical observatory for the observation of the equinoxes, solstices and other events of calendrical significance.

inconceivable. The child psychologist Jean Piaget has shown how consciousness of simultaneity and sequence are *learned* responses in children. Nor are ideas of time universal and uniform. Different languages and cultures have quite distinct ways of representing time. The language of the Hopi Indians of America lacks clearly distinguished tenses for expressing past, present and future. They live in a linguistic perpetual present. Time for them is "what happens when the corn matures or a sheep grows up". Even in Renaissance Italy, Botticelli's painting *The Three Miracles of St Zenobius* depicted three successive moments in time all upon a single stage. What these examples show is that notions of time have a history, and are specific to particular cultures. Our awareness of this is, indeed, a product of *our own* sense of history, living as we do in an age of Relativity. How, then, have notions of time developed?

Human Time

In marked contrast to the modern West, practically all cultures in history, like those of the present "third world", have been small, compact, face-to-face tribal or village communities, occupied primarily with the struggle to win a living from nature. The pace of such traditional cultures has not been dominated by heavy industry, technology and mechanization. The scale, rhythm and measure of their life is a human scale (and by our standards, very leisurely). In such cultures (as still, to a much reduced extent, in ours) time takes on meanings primarily according to human needs.

Thus most societies have had no inkling of, nor would have had any use for, the kind of absolute, linear, uniform, "clock" time which we take for granted. They have been indifferent to accurate, consistent and minute enumeration of time. In peasant societies, people have rarely troubled to remember their own precise age in years. Recording numerical age has become important only with our bureaucratic world with its public registration of births and deaths. Likewise, traditional societies often date important events as being seemingly arbitrary quantities of time in the past. For the *human* significance of time means more than mere numbers.

Thus, attributing very great age to someone (the antediluvian Biblical Patriarchs were supposed to have lived for over 500 years) is a token of their superior wisdom, holiness and venerability. Similarly, time can be reckoned by human, not absolute, measures. The Trobriand Islanders, off New Guinea, date events by saying they occurred "during the childhood of X", or "in the year of the marriage of Y". Many societies have marked history by the regnal year of their rulers. The Romans counted years from the founding of their city. Men have apportioned their own lives not according to their numerical age in years but by the stages of biological and social status: in the time of being a child, a youth, of marriageable

A rendition of Botticelli's "The Three Miracles of St Zenobius", in which three successive events are shown as parts of a single picture.

age, an elder. The periods of life are signalled by rites of passage such as puberty, betrothal and mourning. The human body could be used to measure rate: the pulse, the breath, "in the twinkling of an eye". The length of a day's work or journey might be regarded not as a fixed number of hours but the period until one was tired.

Such ways of allotting time are personal to the individual. But they are also powerfully socially controlled. For time carries public meanings. A man advanced in years would be an elder, qualified for decision-making and the interpretation of law. Law itself is reckoned good because it is ancient. The year has its social rhythms. Time is marked off by festivals, rituals, feasts and fasts, which educate, remind and coordinate the society in its round of work, marking off the times of seed-sowing and harvest, hunting, migration, etc. Rural societies have lucky and unlucky days (one surviving superstition of ours is that Friday the 13th is unlucky). Within Catholic Christianity, Saints' Days, falling at irregular intervals, cultivate the experience of a socially celebrated ritual year. Likewise the sacraments of Christian churches give public religious meaning to the stages of life from cradle (Baptism) to grave (Extreme Unction and Burial). The Seven Ages of Man, spelt out by Jaques in Shakespeare's *As You Like It*, capture in secular form this qualitative, rather than quantitative, measure of life passing.

Of course, the personal and social rhythms of traditional societies must dovetail with natural time ("ecological time"). To take an obvious example, in rural societies the time between dawn and dusk is crucial. For this reason, the Romans had a system of "temporary hours", with a special category of hours of daylight, fixed in number (usually twelve), as also distinct "nocturnal hours" for the hours of darkness. Temporal hours were longer in the summer than in the winter, nocturnal hours longer in winter than in summer, and only at equinoxes were daylight hours the same length as those of the night.

The succession of the seasons is equally important. Of course, what the seasons *are* is experienced differently by (for instance) temperate Europeans (Spring, Summer, Autumn, Winter), Eskimos (who count five seasons), the Nuer of sub-Saharan Africa, who have two main seasons, Wet and Dry, or the Saulteaux Indians of America who reckon six seasons. Civilizations living

A Mayan "almanac" stela, erected in AD497, which serves a two-fold purpose. The elaborate figure in the centre represents a sky god who presides over the five-year period of time inaugurated by the stela, while the many other carvings record the dates and predictions of important events.

close to nature automatically use natural phenomena as a calendar for longer tracts of time. The month is of course a measure of the cycle of the Moon; the year, of the Sun. Some calendars have been highly sophisticated. The calendar of the Mayas of Central America, developed more than a thousand years ago, was in many ways more exact than the modern Gregorian.

Calendars have served practical purposes, such as in agriculture. They also synchronize religious rituals, which themselves mirror the successions of the cosmos. Thus many cultures have performed rituals of the rebirth of the Sun at the time of the winter solstice (remembered within European civilization by the fixing of Christmas at December 25). The Babylonians celebrated a lengthy New Year festival at the Spring equinox, during which the drama of the Creation was re-enacted. Calendars have also served magical and astrological ends. Heavenly bodies, especially when deified, are believed to exercise powers at particular periods over terrestrial affairs. The Maya imagined their gods taking shifts in moving time along, each holding sway during his stint. Compare the twelve astrological signs familiar to Mediterranean civilization; or the association of the seven days of the week with the seven planets, still commemorated in our names for them (Saturday = Saturn Day; Sunday = Sun Day; Monday = Moon Day, etc.). The circle of the signs spells out Fate.

In short, time as conceived in most world communities (the recent history of our own civilization is the major exception) has had two main characteristics:

(a) It has been a measurement of age, duration, and processes by reference to a human yardstick. So it has been relative. "Older than" or "too young", "the first time" or "the ending" are more important than absolute counts of ages. Before and after, or "at the right time", are more eloquent than the precise hour. The time has to be ripe, rather than on the dot.

(b) Time as experience is essentially recurrent and repetitive. It involves cycles of events, of birth and death, growth and decay, mirroring the cycles of Sun, Moon and seasons. The right time to do things comes round again and again at regular intervals.

Transience and Transcendence

Such experiences became distilled in the world's major religions and philosophies. Indeed, religion itself responds to the basic enigma of time: the insecurity for Man of living in the present, conscious of profound past and present dimensions of the Universe

The ancient Egyptian "god of millions of years" who is shown as old and fat as a sign of his great age. The notched stick held in his right hand represents the years of endless time, while his left hand protects an eye that symbolizes the Sun on its nocturnal journeys through the underworld.

over which he has no immediate control, filled with the fear of death and apparent extinction. The solution most religions offer is to stress a mode of Existence which is perpetual, transcendental, eternal, without beginning or end, without threatening and meaningless change: an abode of the gods, or the Buddhist Nirvana.

But religion also integrates the earthly, natural and human present with past and future. The here-and-now race of time thus becomes part of a higher law of continual and endless regeneration. The threat of dissolution is overcome in the idea of endless cycles of time, in which nothing is ever lost or destroyed but all is re-formed and reborn. Thus the Mayan civilization thought time repeated itself in cycles of 260 years. Significant events would follow a preordained pattern. Indian religion believes in the Mahāyuga, the "long year" of 12,000 years, the unit of revolution after which time repeats itself.

Within some faiths, time as a circle, endlessly returning and never destroyed, guarantees rebirth and future life *on Earth*. This is maybe why Upper Palaeolithic men seem generally to have been buried in a crouching posture: perhaps they had been placed in a foetal position in Mother Earth to await rebirth. Belief in the transmigration of souls is central to Hinduism—as it was to Pythagorean philosophy. But, more commonly, the essential function of religion is to overcome the threat of annihilation and the anxieties of the present by assimilating profane Man within the infinite processes of a sacred cosmos. Personal and social decisions (a marriage alliance, a hunt) are made good by ritualistically aligning them with the sacred practice of ancestors, nature and the gods.

THE HISTORY OF TIME

Life becomes (in the words of the cultural anthropologist Mircea Eliade) "the ceaseless repetition of gestures initiated by others". As the Hindu text states: "We must do what the gods did in the beginning." So religious rituals celebrate the New Year by re-enacting the Creation. Other ceremonies perform the triumph of the Sun over Darkness, the coming of the Rains, or the victory of Day over Night. Marriage rituals represent the marriage of Heaven and Earth. Performing a cycle of rituals, repeated scrupulously each year, attunes Man with the cosmos, and overcomes what Eliade has called the "terror of history". Thus the Egyptians ritually buried their dead with images of the god Osiris (who in their mythology died and rose again from the dead), so as to ensure by assimilation the future life of their own dead ones. In the ritual the deceased was given the words: "I am Yesterday, Today and Tomorrow." The Christian calendar annually continues the ritual celebration of the Nativity, Passion, Death and Resurrection of Christ. The Christian Eucharist re-enacts the Last Supper.

Thus religion overcomes the traumas of life in time by assimilating it to a realm of endless time, where time's passing is no threat because it is circular. The profane world of the present becomes adjusted to the sacred world of eternity.

Ancient philosophy was confronted no less by the problem of time. Many philosophers of the Eastern Mediterranean attempted rational accounts of the commonplace experience of time as repetition and recurrence. As Aristotle put it, "Time itself is thought to be a circle". Such an idea was common amongst the Romans, too, as well as the Greeks. Thus for Seneca, "all things are connected in a sort of circle. Night is close at the heels of day; day at the heels of night; Summer ends in Autumn; Winter rushes after Autumn, and Winter softens into Spring . . . all nature in this way passes, only to return". Some four hundred years earlier, Plato believed that the succession of years was programmed to repeat itself at a fixed interval, that of the Great Year, which would last 36,000 solar years. The Pythagorean philosophers likewise took the view that "everything will eventually return in the self-same numerical order", and Aristotle's followers speculated whether, in this system of endless return, Paris would once more carry off Helen, sparking off another Trojan War.

Seeing time as a circle destroyed the threat posed by time the destroyer. Classical philosophy was deeply troubled by the everyday passage of time and events. Thinkers like Heraclitus could see in the here-and-now world of time nothing other than

12

meaningless chaos and hectic flux. Time spelt change, and change meant decay and disintegration. Thus the poet Hesiod believed that the first men had lived in a blessed Golden Age, when the Earth was freely productive, life was easier, and men were virtuous and did not need to labour. Since those days society had corrupted through a Silver Age into an Iron Age. Plato thought government went through inevitable decay from the rule of philosopher kings down to tyranny. The historian Polybius saw in politics nothing other than a dismal circle of ceaseless instability and revolutions.

Classical philosophy found two remedies against change and decay. One was to construe time as an endless cycle. For the cyclical was perfect. The circle came back to its own starting point. It had no "loose ends". It eliminated the problems of a beginning and an end to all things by postulating the infinite duration of the cosmos. The other was to suppose a level of reality which was immune from change: the realm of eternity. This was the transcendental plane untouched by worldly matter, that of Ideas. In the vision of the Pythagoreans and Plato, the highest reality consisted of ideal forms (which were timeless, though they could be conceived spatially), such as the Idea of the Good or the Idea of Perfect Geometry. It was precisely because this intellectual world was timeless (therefore, unchanging) that it could be known. The world-in-time was at best a poor imitation or substitute for this ideal Eternity—or, in Plato's evocative phrase, no more than "the moving image of eternity", where, by "moving", he meant "imperfect".

God and Time: Judaism and Christianity

Our own culture is permeated by such views of time as cyclical, and of change as decay. Witness the idea of the wheel of fortune, or of the rise and fall of civilizations, or the ambiguities of the word "revolution". But our vision of time has been chiefly shaped by the most stunning *exception* to the view of time as eternal succession: the outlook of the Jews as absorbed and developed within Christianity. Through their long history of tribulation, exile and persecution, the Jews developed faith in Yahweh—a unique, personal, presiding God, utterly superior to the trumped-up nature deities of other tribes. They were His chosen people. Hence Jewish religion—and, later, Christianity—exceptionally saw God as the *Creator* of the entire Universe. As the first verse of the first chapter of *Genesis* runs: "In the beginning God created the heaven and the earth" (which was later glossed as meaning "created out of nothing").

An illumination from a fifteenth-century French manuscript depicting God the Creator receiving the soul of a dying man. Christians saw time no longer in terms of repeated cycles of events, but as linear and finite, created by God and to be brought to an end by him.

Furthermore, God the Creator of the Universe would also be its destroyer, or transformer. As the last chapters of the last book of the Christian scriptures, the *Book of Revelation*, prophesied: "I saw a new heaven and a new earth, for the first heaven and the first earth were passed away"; for the Lord said, "I am Alpha and Omega, the beginning and the end, the first and the last".

Thus time was no longer deified. The circles of time no longer set the pattern and measure for life and nature. Time itself was dethroned. Time was now a creation of God. Time was finite, God infinite. Hence time could not apply to God Himself. In St Augustine's words, "God, whose eternity alters not, created the world in time"; "the world was made with time and not in time". Though eternity became a mystery, time itself lost its threat. As Sir Thomas Browne explained in his *Religio medici* (1635), "Time we may comprehend; 'tis but five days older than ourselves".

The belief that God had created it revolutionized the understanding of time itself. Time ceased to be endless, repeated cycles of events, and became linear, sequential, unique, irreversible. Time was teleological. Man's fate—the pilgrim's progress—had been mapped out by a predestining God and revealed in advance by prophecy. "A straight line traces the course of humanity from initial fall to final redemption." Christ had died for Man's sins. The notion of an endless cycle of Christ atoning over and over again for Man's sins was grotesque. Within Judaeo-Christianity, a scale of absolute time became possible, because years could be counted forward from the Creation. History was God's design unfolding over time, God's will in the world. "We looke upon God, in History," wrote John Donne, "in matter of fact, upon things done, and set before our eyes; and so that Majesty, and that holy amazement, is more to us than ever it was to any other Religion." A complete genealogical lineage could be traced forward from the first man, Adam, up to the present generation. Adam's sin had infected them all.

Time could be seen as a series of stages, leading from the Creation up to the prophesied end of the world, when Salvation would be Christianity's answer to the "terror of history". St Augustine traced six ages up to and including the present, emblematically representing the six days of Creation as recorded in *Genesis*: (1) from Adam to Noah, (2) from Noah to Abraham, (3) from Abraham to David, (4) the Babylonic Captivity, (5) from the Babylonic Captivity up to the Incarnation, and (6) from the Incarnation to the present. A seventh age was to come, that of Man's heavenly rest with God (of which the seventh day of Creation was a token).

This linear progression could be given numerical form. The six ages added up to about 6,000 years' duration (Psalm 90 stated that a thousand ages are but a day in God's sight). Hence God would bring the present world to an end 6,000 years after the original Creation (whether this event would be the New Jerusalem, the Second Coming, or the Last Judgment itself, was subject to debate). The history of Christianity is littered with claims setting the date of the Divine Intervention. The year AD500 was an early favourite; then the so-called year of the Millennium, 1000. The monk Joachim of Fiore (*c.*1132–1202) predicted 1260; many English Puritans marked 1666 as the year of the Millennium. Despite the failure of these predictions, this basically linear viewpoint has shaped the perception of time in Christian civilization ever since. The New Testament follows the Old. We divide our time into BC (Before Christ) and AD (*Anno Domini*: in the year of our Lord).

Christians have always deprecated the world-in-time in contrast to that of eternity. The world-in-time contains death, decay, the

The traditional figure of Old Father Time. Note the hourglass on his head, indicating the relentless passage of the "sands of time", and the scythe with which lives are cut short.

futile pomp and bustle of scurrying men. In Medieval icono-
graphy, the figure of Death was often pictured carrying an hour
glass. Time was the servant of Mortality. Father Time is shown
carrying the sickle of destruction. Early clocks bore epigraphs like
"*Tempus fugit*" (Time flies), "*Mors certa, hora incerta*" (Death is
certain, life uncertain), "*Toutes les heures vous tue*" (The hours are
killing you, every one of them) or, later, "Time and tide wait for no
man". Time was thus seen as a ravager. Shakespeare wrote of
"mis-shapen time", "the bloody tyrant time" and "time's in-
jurious hand". Life was "time's fool". Many of his sonnets struggle
with the wasting effects of time, a mood reinforced by the
Reformation belief in an imminent end to earthly affairs. Luther
prophesied that "the world will perish shortly; the last day is at the
door, and I believe the world will not endure a hundred years".
Such views could still be echoed over a century later by Sir
Thomas Browne: "The world grows near its end."

The Christian sense of the brevity of time harrowed the spirit.
Thus Andrew Marvell appealed *To His Coy Mistress*:

> Had we but World enough and Time
> This coyness, Lady, were no crime.

He would like to woo her slowly, but dared not because

> . . . at my back I always hear
> Time's winged Chariot hurrying near;
> And yonder all before us lie
> Deserts of vast Eternity

—that is, Death. Now

> The Grave's a fine and private place,
> But none, I think, do there embrace.

Because time was so fleeting, the poet Herrick advised buoyant-
ly: "Gather ye rosebuds while ye may." But other Christian writers
found the world-in-time nauseating and irksome. Thus William
Blake:

> Ah, Sunflower! weary of time
> Who countest the steps of the Sun,
> Seeking after that sweet golden clime
> Where the traveller's journey is done,

Where the Youth pined away with desire,
And the pale Virgin shrouded in snow
Arise from their graves, and aspire
Where my Sun-flower wishes to go.

No wonder the blessedness sought by Christian mystics has been that of the obliteration of earthly time. As the late Medieval mystic Meister Eckhardt phrased it, "there is no greater obstacle to Union with God than Time". So Christians have possessed a different solution from that of pagans to the "terror of history", but they have felt no more comfortable about Man's existential life in time.

Time as History

All societies possess some conception of time past, and of their own ancestry. This can take many forms. Lists of dynasties; the genealogies of kings and the pedigrees of aristocrats; annals recording the events of each year; monuments celebrating great victories; myths, chronicles and epic tales of heroes and ancestors ("once upon a time . . ."); or religious accounts of origins interpreted as the deeds of the gods. Societies need such "usable histories". They make sense of the present in terms of the past; they back up the traditional power of rulers; they give tribal or patriotic identity; and they establish what is good in moral codes, the law, or religious practice by reference to what is time-hallowed.

But this is very distant from our current sense of history. To us, the memories and myths of most cultures play fast and loose with exact chronology. They are uncritical in their use of evidence. They muddle legend and history, men, gods and heroes, the factual and the fictional, truth and literary effect. This is hardly surprising. Most societies possess scant—if any—written documents and objective records authenticating their own past. The past fortifies the present, rather than being the object of impartial, detached inquiry. Because most societies are static and deeply conservative, past and present tend to merge in a confused haze. The past does not have a distinct identity.

How little of their own history even the Greeks knew! The Greeks had short memories. Even the events of the Trojan War (*c.*1250BC) were beyond historical inquiry, and were merely the subject of legend, and romance, such as Homer's. The best Greek history—such as Thucydides' history of the Peloponnesian War—was contemporary history. The same is true of Roman historians like Caesar, Sallust and Tacitus. Livy's history of Rome, by

Images of the "terror of history": the hourglass, the skulls of the dead, and the sleeping child.

contrast, lacked documentation for the early years of the city's history; he did little more than transpose back the political myths and prejudices of the late Republic. In any case, the aim of most historians in the Graeco-Roman world was to teach politics by concrete examples. Believing that human society moved in cycles and that human nature was constant, historians like Polybius used the past to drive home lessons for the present.

Within Christendom, belief in the pre-ordained course of human destiny from Creation through to the Last Judgment gave special importance to history, for every human act possessed a providential place in the Divine time-table. But through the Middle Ages Christian history remained preoccupied with miracles and marvels. Showing how the finger of God in history blessed the pious and punished the infidel, it served the purpose of edification, particularly in writing the lives of the saints (hagiography). Low standards of scholarship were matched by high levels of credulity.

The roots of the modern concept of history lie in the age of the Renaissance and Reformation. The rediscovery of manuscripts during the Renaissance brought new evidence to light; printing made it more readily available. Scholarship improved. New critical methods were devised, such as the techniques of philology (the study of words) and diplomatic (the science of official forms), to date and authenticate documents. Renaissance scholars like Lorenzo Valla used such tools to expose forgeries perpetrated by the Medieval Papacy. Approaches to history also became more secular; human causation took precedence over divine intervention.

But, above all, European scholars began to recognize for the first time just how *distinct* their own times were from former days. Renaissance men admired the societies of Classical Greece and Rome, and had hoped to imitate them; but, the more they studied such cultures, the more they found that they were essentially inimitable. For modern Europe possessed hitherto unknown technological advances—like printing, gunpowder and the compass. From 1492 modern society knew about the new continent, America. Similarly, Renaissance men found that modern society was in its legal principles fundamentally different from the Roman Empire regulated by the Justinian Code. Thus Roman law had dealt extensively with slavery, unknown to sixteenth-century Europe, but had nothing to say about the feudal tenures and commercial transactions so common to the modern economy. Early Protestants found the practices of contemporary Christians far different from the worship of the pristine Christian churches.

This discovery of essential differences between past and present times is the kernel of modern historical inquiry. It produces for the first time an awareness of "anachronism". Understanding the congruence between particular institutions, ideas or laws, and their precise chronological time became the task of Western historians. The love of the past for its own sake—what we might call "antiquarianism"—came into being. Scholars were fired with the desire to collect all possible documents and remains of the past—inscriptions, works of art, pottery, coins, and so on. Since the seventeenth century the main goal of scholarly history has been to find out "what really happened" (the words of the nineteenth-century German historian, Ranke: *wie es eigentlich gewesen ist*), above all on its own terms. Each person, each event, each period is to be treated as equal and autonomous, significant for its own sake. Each historical moment has its own meaning in itself; each age its own spirit, its *Zeitgeist* (the "spirit of the age").

In the sixteenth and seventeenth centuries, men were generally pessimistic about the distinct nature of their own times. Man had begun innocent and virtuous. The "Ancients" (by which term was meant the Greeks and Romans) had been better poets, philosophers and scientists than the "Moderns", because they had lived in the good old days, when the human mind was freshest, most imaginative, uncorrupted by the accumulation of errors. When the early Renaissance poet Petrarch called his own times the "New Age", his word "new" had a pejorative ring. Change was still decay. What was old still possessed authority.

History and Progress

But change was afoot. For, more and more, when people compared their own times with Classical Antiquity and with what became known as the Middle Ages, they felt pride at the improvements which had taken place. Aside from advances like printing and the discovery of America, modern scientists like Copernicus, Kepler, Francis Bacon and Newton were exposing the ancient errors of Aristotle and Ptolemy. "This is the age," wrote the seventeenth-century English scientist Henry Power, "wherein all men's Souls are in a kind of fermentation . . . Philosophy . . . comes in with a Spring-tide." Modern poets and playwrights like Shakespeare, Cervantes, Corneille and Racine could stand shoulder to shoulder with Homer, Aeschylus and Virgil. The seventeenth century judged the relative merits of the Ancients and Moderns, and concluded that the story of the sciences and arts was indeed the story of progress. What was "new" could now, for the first time, be seen not as decadent or presumptuous but rather as original, advanced and forward-looking. Thus Kepler could boast of his "New Astronomy", Galileo of the "Two New Sciences" and Francis Bacon, as part of his plans for the "advancement of learning", could sketch a perfect, Utopian society called the "New Atlantis", superseding the old Atlantis of Plato.

These developments—what the sixteenth-century physician Jean Fernel called "the triumph of our New Age"—altered perceptions of the place of the present in time. Gradually the present in respect to the past came to be seen as the climax of a long series of stages of development. Elaborating this, speculative philosophical historians of the eighteenth and nineteenth centuries like John Millar, Condorcet, Herder, St. Simon and Auguste Comte developed sweeping optimistic visions of the evolution of the human mind and society. Man had moved from an isolated state of nature to organized social life; from rudeness to refinement; from savagery to civilization; from barbarity to humanity, from selfishness to benevolence. Toil and poverty were being alleviated by labour-saving technology and the growth of wealth, as economies advanced from nomadism to agriculture, and from agriculture to commerce and industry. Ignorance was yielding to knowledge, irrationality to science. Francis Bacon thought that truth was the daughter of time. He pictured the ship of knowledge sailing out through the pillars of Hercules from the closed world of the Mediterranean "*plus ultra*" (yet further) into the new limitless ocean of the Atlantic, advancing "the effecting of all things possible". The primitive religions of the ancient world, with their polytheism, superstitious animistic cults and fear of the unknown, had been replaced by worship of a single, rational, all-wise, benevolent, merciful God (or, thought the daring, religion had been replaced altogether by Man's emancipation into science and self-dependency). And, gradually, political tyranny was yielding to representative government, freedom and democracy.

Hence, from the eighteenth-century Enlightenment, a new time-perspective on mankind triumphed, which saw history as the history of progress. At last time seemed on Man's side—though crusty reactionaries like Dr Johnson complained that "the age is running mad after innovation". "We are on the side of progress," portentously pronounced the nineteenth-century historian Thomas Macaulay. There was a new faith in Man's ability to be master of his own fortunes. What made progress possible? For the German, Herder, it was because, unlike other animals, Man had an infinite capacity to learn, to educate himself. For Hegel (1770-1831) progress was possible because consciousness continually expanded. For others, science was the key. Echoing Bacon, Macaulay argued that science ensured the "great and constant progress" of mankind: "It has lengthened life; it has mitigated pain; it has extinguished diseases . . . it is a philosophy which never rests, which has never attained, which is never perfect. Its law is progress."

For others the secret was economic advance. Thus for Karl Marx progress was assured (though dialectical) because Man could find ever more efficient ways of exploiting nature and appropriating its fruits for his own benefit.

Equating the passing of time with progress had another profound effect. It transformed expectations for the future. From the time of the Apostles down to the end of the seventeenth century, many Christians had believed the end of the world to be nigh. The current age was thought to be full of irremediable evils.

(*Above right*) Sir Thomas More (1478–1535), an engraving by R. Woodman from an enamel after Holbein. More is best known for his philosophical exploration of an ideal society, *Utopia*. (*Below right*) The frontispiece to *Utopia* (1516).

In dreaming of a perfect society, men had envisaged it either in the past (Paradise, the Garden of Eden, the Golden Age), or as outside the course of human history altogether; hence the Christian notion of Heaven, or the idea of "Utopia", originally put forward by Thomas More in 1516 and embellished by others. Utopia was an ideal society existing only in the mind. For More the word "Utopia" meant nowhere.

But from the eighteenth century men turned their eyes towards the future. Western capitalism was multiplying its wealth. It was conquering the rest of the globe. Medicine was controlling epidemic disease, improving life's comforts, and raising life expectancy, at least for the more affluent members of society. Hence, the future could be looked to with confidence and eagerness. The decline of belief in the Millennium meant that society seemed to possess an indefinite future span in which to continue developing. Man could make himself. He had progressed, but was not yet perfect. Indeed, the very notion of final perfection gave way to that of perfectibility—endless and continued improvability. "The perfectibility of man," wrote Condorcet, "is truly infinite." The Industrial Revolution was the final morale-booster; the railway, for example, suggesting Tennyson's

Not in vain the distance beacons. Forward, forward let us range,
Let the great world spin for ever down the ringing grooves of change.

So the previous vision of Paradise Lost, or a paradise to be regained in heaven, became translated into remarkably secular terms: a future paradise on Earth to be created by Man's efforts. In the words of Joseph Priestley: "Knowledge, as Lord Bacon observes, being power, the human powers will in fact be enlarged; nature, including both its materials and its laws, will be more at our command; men will make their situation in this world abundantly more easy and comfortable; they will probably prolong their existence in it, and will grow daily more happy, each in himself and more able (and I believe more disposed) to communicate happiness to others. Thus whatever was the beginning of this world, the end will be glorious and paradisiacal beyond what our imaginations can now conceive." Thus the Western capitalist world has no time now for what is old-fashioned or past or obsolete. Industry builds obsolescence into its products to ensure future change. We have a cult of the new.

Turner's "Rain, Steam and Speed" expresses the nineteenth century's pride in the technological revolution.

Obviously the last two centuries have produced also profoundly pessimistic predictions of the future, predictions which have feared industrialism, towns, bureaucracy. In the early nineteenth century Romanticism protested against the mechanization of the human spirit. The "counter-culture" of the 1960s rebelled against a future dominated by technology and money values. Conservation movements believe that modern science and technology are jeopardizing—not improving—Man's future. Aldous Huxley's *Brave New World* and George Orwell's *Nineteen Eighty-four* warn of Man's power to manipulate and control his fellows.

The point is, however, that, whether the vision is optimistic or pessimistic, the orientation of the vision has been transformed. In the original Christian outlook, the total lifespan of the human race was to be a mere few thousand years. Man was a "given" entity, with a more or less fixed nature (he had been created in God's image). Hence Man's understanding of himself was essentially static. By contrast, over about the last three hundred years, a profoundly historical perspective has developed. It is the view that Man has gradually evolved from the primitive state to civilization, and that the future stretches equally far beyond. In Ortega y Gasset's famous phrase, Man has no nature, only a history. Man is what he has become. Man has made himself. He has a limitless capacity to change himself. Man is the product of time.

From Natural History to the History of Nature

As notions of Man in time were changing, so were views of nature. Just as the vision of Man moved from a static one to one radically historical, so the understanding of nature was also historicized.

In the Judaeo-Christian view of nature, the cosmos was created immediately before Man, less than 5,000 years ago. Although details were subject to great dispute, the essence of this belief still commanded support at the end of the seventeenth century. What the Biblical account seemed to emphasize most was that Creation was all of a piece. God had created the heaven and the Earth, then light, had separated land and sea, had produced the Sun and Moon, and then created vegetable and animal forms on Earth, culminating in Man, all in a week of creative energy. God had seen that it was good and complete, and then stopped. Because all things were seen by God to be good as created, there was no need for tampering and improvements. The Bible made no mention of subsequent new creations. And nothing that God had made could have ceased to exist. The point of the story of Noah's Ark was that

couples from every species of animal had been preserved to populate the Earth afresh after the Flood had wrought God's punishment on disobedient Man.

To put this another way: Medieval and Renaissance naturalists were confronted with the problem of understanding various aspects of their environment—how to classify the relationships of different creatures to each other; why flora and fauna were distributed the way they were; the arrangement of land and sea and the landscape. They did not anticipate that the dimension of time had much part to play in answers to these questions. For the globe was basically as God had intended it to be. There certainly had not been enough time for great change and, in any case, most alterations (like the Flood) were attributable to specific, Divine, interventions. Thus, to understand Nature was to understand not its history—for it hardly had a significant history—but rather its *design*: why God had made things the way He had.

The Creation, as envisaged by Michelangelo in the ceiling to the Sistine Chapel. In the traditional view, Man—epitomized by Adam, seen receiving the "spark of life" from God—was created on the seventh day; it is interesting to note that, if the word "period" is substituted for "day", the chronological account given in *Genesis* is approximately correct.

In other words, science traditionally saw the order of nature in terms of relationships which were atemporal. These might be analogical. The Earth might be compared to the human body, and understood by comparing the functions of rivers, mountains, volcanoes with corresponding facets of the human physiology and anatomy. Both the body of Man and that of the Earth were composed—thought Aristotle—of elements and humours. Gems and metals were believed to correspond to the planets. Or the relationships might be purposive; thus the nature of vegetables was to provide food for animals, and of animals to provide food, and to serve as beasts of burden, for Man. For God had commanded Adam to have "dominion over the fish of the sea, and over the fowl of the air, and over the cattle, and over all the earth, and over every creeping thing that creepeth upon the earth" (*Genesis*: 1: 26).

From about the sixteenth century, this vision of a static, recent, unchanging Universe began to be challenged. There are many reasons for this. It happened partly because scientists successively came upon new evidence—such as "novae" ("new stars") or fossils—which could not easily be assimilated within the old picture. It was partly because, particularly from the eighteenth century, many scientists had themselves ceased to be Christian believers, or at least were taking the view that religious faith should not be mixed up with scientific investigation. Also, as Western Man's own vision of himself in society became transformed by ideas of time and history, it became more attractive to think of nature in similar terms.

The new heliocentric astronomy of Copernicus, Kepler and Galileo made the first major break, for it removed the Earth from the centre of Creation: Earth was demoted to being merely a minor planet circling just one star in a Universe which Descartes and Newton were declaring was infinite in space. The idea of a Universe of infinite space did not immediately lead to the idea of infinite time (although it obviously made the enlargement of the time dimension easier to stomach when it did come); it did, however, suggest the idea that the *Genesis* account applied not to the creation of the whole of the cosmos, but only to that of the Earth. Thus other parts of the Universe might have antedated the Earth by incalculable quantities of time. Similarly, cosmologists like Thomas Burnet and Newton's protégé William Whiston interpreted the Biblical "Last Things" as though they applied not to the destruction of the entire Universe, but only to that of the Earth. Thus it became for the first time admissible that other worlds predated the Earth. The respectable French philosopher Fontenelle could indicate, in his *Conversations on the Plurality of Worlds* (1686), that other stellar systems contained intelligent life, like men, presumably operating with timescales different from that of the Earth.

Nevertheless, most seventeenth-century naturalists continued to believe that the *Genesis* timescale applied more or less literally to the Earth. They were anxious lest, if Moses (the presumed author of *Genesis*) were mistaken about Creation, then the Decalogue might be thought so too. People were also afraid that allowing a much greater antiquity to the Earth was the thin end of the wedge. The doctrine of the eternity of the Earth would follow, which would spell atheism and destroy Man's hopes for salvation. Moreover, great draughts of time did not seem necessary. For if, indeed, God had miraculously intervened to change the Earth (e.g., the expulsion of Adam and Eve from Paradise), problematic phenomena, such as what seemed like organic remains deeply embedded within the rocks (i.e., fossils), could be accommodated within the Biblical timescale. Nevertheless, the Biblical chronology was already on the defensive. Meditating upon fossil ferns of a type not found in the contemporary world, John Ray in the seventeenth century was afraid that "there flows such a train of consequences, as seem to shock the Scripture-history of ye novity of the World; at least they overthrow the opinion generally received & not without good reason, among Divines and Philosophers, that since ye first Creation there have been no species of Animals or Vegetables lost, no new ones produced". Already other naturalists, however, faced by evidence of the reversal of the positions of land and sea during Earth history, felt obliged to concede that the Earth was indeed older than was commonly assumed—although practically no one was prepared to hazard a date for its origin.

The major shift was one of attitudes. Scientists increasingly adopted the stance that, for explanations of the configuration of the world to be satisfactory, they must be in terms of the regular operations of nature, through routine laws, excluding miraculous interventions. The first important explanation of the position of the heavenly bodies—in terms not of where God had originally placed them, but rather how they had naturally come to be there—was the so-called "nebular hypothesis", proposed by Kant and later by Laplace. This, when expanded, argued that the "fixed stars" were not fixed after all. The Universe had begun as a chaotic cloud of swirling gases, which in course of time had organized itself, under its own gravitation, condensed in parts, rarified in others, and formed into the various stellar groupings. Similar attempts to explain the figure of the Earth as having developed from a chaotic, unorganized original mass dated back at least to Descartes' *Principles of Philosophy*. However, for fear of running into ecclesiastical opposition as Galileo had done, Descartes had emphasized that his was not an account of how God had actually programmed the development of the Earth, but rather just one way God might have done so had he wished. Leibniz in his *Protogaea* similarly saw the Earth as having evolved out of an original fiery mass by processes of cooling, separation and solidification.

What is important about all these views is that they sought to explain the present structure and configuration of the Earth in terms of a historical development which was gradual, steady, uniform, and progressive; and so, by implication, slow and lengthy. The meaning of the Earth lay not in its original design, but rather in what it had become. Creation was not a once-and-for-all act, but rather a continuing emergent process. As Kant said, "Creation is not the work of a moment."

In the eighteenth and nineteenth centuries this approach was extended in two complementary ways. Firstly, attempts were made to determine the age of the Earth. Buffon's was the pioneering attempt: he tried to infer the period for the cooling of the Earth by experimental analogy with the cooling of iron balls. In public he suggested a timescale for the Earth of some 74,000 years down to the present; privately, he speculated that the Earth might have a lifespan of perhaps half a million years. The stumbling-block for all such attempts lay in finding an objective, current measure of the rate of terrestrial change. In 1715 Edmond Halley had suggested computing back from the rate of the increasing salinity of the oceans, but this proved impracticable. Other attempts to find such "natural chronometers" were highly unreliable. Nevertheless, by the 1830s reputable geologists were measuring Earth history in millions rather than thousands of years. In a rather unguarded remark, Charles Darwin hypothesized that the Wealden Greensand formations of South Eastern England might have been formed no less than 300 million years ago—generous even by today's standards!

Such a timescale, of course, utterly destroyed the literal reading of *Genesis*. This greatly disturbed some Christians. Thus the art-critic John Ruskin wrote in 1851: "If only the Geologists would let me alone, I could do very well, but those dreadful Hammers! I

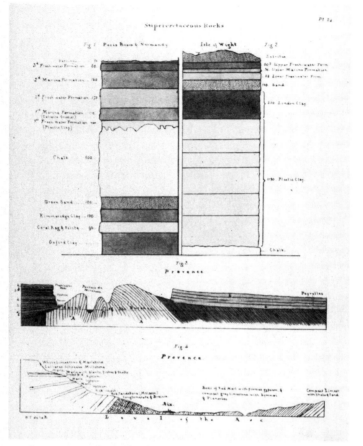

A nineteenth-century geological column showing the vertical order of rocks with the presupposition that it also represents a time order.

hear the clink of them at the end of every cadence of the Bible verses.'' But most Christian naturalists could accommodate these discoveries by stating that *Genesis* referred, after all, only to the age of *Man's* origin. The main challenge to the geologists' dating of the Earth came, in fact, from fellow scientists. Using modern thermodynamics, the physicist William Thomson (later Lord Kelvin) argued that, if the Earth had been flung off the Sun and had cooled normally, not more than about 60 million years could have elapsed since it had been molten. However, a newer physics eventually rescued the geologists with their commitment to the extreme gradualness of geological change. The discovery of radioactive processes around the turn of the twentieth century by Henri Becquerel and Ernest Rutherford showed there was a hitherto unknown heat source in the Earth, heat being emitted from minerals at a very slow rate. Geologists currently believe that the Earth first acquired a permanent crust some 2,800 million years ago, although rocks over 3,000 million years old are known. The earliest known rocks, the Precambrian, date from over 600 million years back, the Secondary System from about 350 million years, and the Tertiary from about 50 million; the last few million years are described as the Quaternary.

The second way of demonstrating that the story of the Earth was one of gradual historical development was by showing how commonplace geological processes (such as the action of frost or rivers) would produce immeasurably large consequences, *given enough time*. This principle of the "parsimony of force and the prodigality of time" in nature is associated with geologists like James Hutton and Charles Lyell. Hutton argued that we have no ground for limiting the amount of time available to work geological changes, for we have "no vestige of a beginning,—no prospect of an end" to the Earth's system. The Earth was thus indefinitely old. In this "Uniformitarian" school of geology, time was conceived as a natural force in its own right. Using a mathematical image, Hutton's friend Playfair wrote: "Time performs the office of integrating the infinitesimal parts of which [the progress of the Earth] is made up." Buffon similarly wrote: "The great workman of nature is time. He marches ever with an even pace and does nothing by leaps and bounds; but by degrees, gradations, and succession . . . ; the changes which he works are first imperceptible; become little by little perceptible, and show themselves eventually in results about which there can be no mistake." The early Victorian geologist George Poulett Scrope summed up the historicization of the Earth: "The leading idea which is present in all our researches, and which accompanies every fresh observation, the sound which to the ear of every student of nature seems continually echoed from every part of her works is

'—Time! —Time! —Time!'"

The History of Life

Once the Earth had acquired a history, the same arguments were bound to be applied to the organisms which inhabited it. Up to the eighteenth century the existence of distinct but morphologically comparable species of plants and animals had been understood within the framework of a preordained, timeless, static, hierarchic Divine plan, the Great Chain of Being. On this every species had a place, from the lowest form of plant life up to Man. Each species was assumed to have come into being by Divine *fiat* at the Creation. Species were basically fixed. The task of the natural historian, like Linnaeus, was to classify life into classes, genera, species and varieties. Certain naturalists felt dissatisfied with his theological account of the differentiation of species; but traditional alternative accounts of the *natural* origin of species (e.g., that they had been spontaneously generated out of some primeval slime) won little scientific support. And early theories of evolution—such as those of Lamarck and Erasmus Darwin (Charles Darwin's grandfather)—were dismissed as too speculative.

The traditional static vision of the order of life began to crumble from about the time of the French naturalist Cuvier, early in the nineteenth century. The decisive discovery was that quite distinct populations of organic creatures were to be found fossilized in the strata of successive geological epochs. Many of the earlier species had undoubtedly become extinct—most spectacularly the giant pachyderms and dinosaurs. There seemed to have been a rough temporal progression of life from relatively simple, invertebrate forms (bivalves, brachiopods, etc.) found in early beds, up through fish, reptiles and primitive mammals (marsupials) to the higher mammals, primates, and finally Man. This proved that life had a history—a progressive and immensely long one. It did *not* prove, however, that life had "evolved". Indeed, most palaeontologists argued that populations of species had been created by God in successive bouts of special creation, one after another. Yet the *extinction* of species was being accepted as due to natural causes (e.g., competition), as also their geographical distribution.

From about the 1840s, however, evolution was in the air again, in the writings of Herbert Spencer and Robert Chambers, who deemed that "the inorganic has one final comprehensive law, Gravitation. The organic, the other great development of mundane things, rests in like manner on one law, and that is Development." It was Charles Darwin, in his *Origin of Species* (1859), who marshalled the evidence to convince the scientific world, not just that life had a history, but that new species organically evolved out of previous ones. Darwin's mechanism for evolution was that chance variations helped certain creatures to survive better under conditions of environmental competition: the principle of natural selection (or, in Spencer's phrase, "the survival of the fittest"). Darwin's mechanism did not win the general support of naturalists until the 1930s.

There was greater resistance to seeing Man in this perspective.

By about 1830 scientists were willing to interpret the Universe, the Earth and life itself as being incalculably old, and as the product of time in history. Yet they were convinced that Man must be an exception. Man (it was claimed) had existed for but a few thousand years (the *Genesis* timescale), and had been specially created in God's image. Nevertheless, from anthropologists' studies, the diffusion of race, language and culture seemed to require a longer period of development. Archaeological evidence, from peat-bogs, caves and gravels, was showing that human remains (e.g., flint tools) were coeval with geological beds tens of thousands of years old, and even with extinct animals. The Dane Christian Thomsen's periodization of Man's early history into a Stone Age, a Bronze Age and an Iron Age had become established by the mid-century, bringing with it the ideas of "prehistory" and "social evolution". Furthermore, bones which seemed to be transitional forms between the higher apes and Man were coming to light: "Neanderthal man" in Germany in 1856; Cro-Magnon man in the 1870s. By the 1860s the educated public was beginning to accept that Man was a creature of high antiquity. But Darwin's account of organic human evolution in his *The Descent of Man* (1871) still met enormous hostility.

When the evolution of Man was finally accepted in the latter part of the century, most scientists, philosophers and theologians continued to contend that certain elements of Man were exempt from the law of natural development in time. Thus the mid-twentieth-century Papal ruling that the "doctrine of evolution" should be investigated by scientists "in so far as it deals with research on the origin of the human *body*"; nevertheless "the Catholic faith obliges us to believe that *souls* were created directly by God". Twentieth-century anthropology, particularly in the researches of Raymond Dart and Louis Leakey and his family in Africa, has traced back a succession of ever more primitive hominid (man-ape) types, such as *Australopithecus*, to about three million years.

Thus views of the Universe, the Earth, Life and Man were all historicized. Time was directional, for evolution was irreversible. This vision chimed well with the implications of the Second Law of Thermodynamics as formulated in the mid-nineteenth century. Physicists argued that the material world was tending to disorganization (entropy); or, to put it another way, that the amount of usable energy in the Universe tends to diminish. Both entropy and evolution showed time to be directional; to be arrow-like.

Darwin himself expressed optimism about evolution: "As natural selection works solely by and for the good of each being, all corporeal and mental environments will tend to progress towards perfection." Yet writers and artists responded to the evolutionary view that Man was just a higher primate who had developed late in geological time in a Universe which would suffer a thermodynamic "heat death" with a sense of loss, emptiness and purposelessness. Thus Tennyson's agonized question in *In Memoriam* (1850):

> . . . *and he, shall he,*
> *Man, her last work, who seem'd so fair* . . .
> *Who trusted God was love indeed*
> *And love creation's final law—*
> *Tho' nature red in tooth and claw*
> *With rapine, shriek'd against his creed* . . .
> *Be blown about the desert dust*
> *Or seal'd within the iron hills.*

Ironically, the earlier terror at the fleetingness of time had turned into the absurdity of meaningless, impersonal eons.

Objective Time

In the physics of Aristotle, as passed down by the Medieval scholastics, time was conceived as a measure or function of motion. Time was relative to actual bodily movements, to "becoming". During the Scientific Revolution of the seventeenth century this relationship was reversed. Time now came to be construed as a universal, background dimension, against which other physical properties (such as motion) could be measured. Time was separated from its physical content. In the words of Isaac Barrow, "whether things run or stand still, whether we sleep or wake, time flows in its even tenor".

The view was formulated that time was a dimension in its own right: objective, universal, and abstract, one axis of a grid of nature (the other being space) upon which every object and motion could be plotted. The classic definition of this view was given by Isaac Newton at the beginning of his *Mathematical Principles of Natural Philosophy* (1687): "Absolute, true and mathematical time, of itself and from its own nature, flows equably without relation to anything external." Such a conception was of paramount importance to the new, abstract, quantified mechanical philosophy. Time, thus conceived as an ideal, absolute standard, permitted mathematical calculations of velocity, acceleration, duration, etc. As Newton's own teacher, Isaac Barrow, wrote: "Time implies motion to be measurable." Absolute time could thus become capable of geometrical and mathematical handling in formulae and equations. In the words of Newton's friend, John Locke, "Duration is but as it were the length of one straight line extended *ad infinitum*." This view, that there was one single standard measure of time through the Universe, served physicists and cosmologists well until the present century: in Chapter 5 the modern physicist's view of time will be fully discussed.

Perhaps the greatest stimulus to seeing time as an objective, absolute, background dimension was the development of the clock, which will be examined more fully in Chapter 3. Mechanical clocks, propelled by falling weights and regulated by escapement mechanisms, had been constructed in the West from the thirteenth century. Public clocks (which struck the hours but did not have a face with hands) appeared in Italian towns from the fourteenth century. By the next century both domestic clocks and the alarm clock had come into existence. The theoretical division of hours into minutes and seconds, and of the day into a.m. and p.m., date from about this time. From the seventeenth century, well-off people could afford pocket-watches (Samuel Pepys records in his diary how proud of his he was: he walked around telling strangers the time). Early clocks were not very accurate but, with the invention of the pendulum by Galileo and Huygens in the seventeenth century, clocks became accurate to within about ten seconds a day.

The clock revolutionized Man's sense of time. The subjective reckoning of time—judging time by stints of work, by tiredness, by "longer" and "shorter"—gave way to the relentless, steady, linear, uniform, objective ticking of clockwork. Henceforth time would be judged as "such-and-such o'clock". Previous measures of time, like the pulse, the sundial or the waterclock, had relied on the rhythms of Man or nature—some days the Sun did not shine; in winter the water froze. Now the clock could itself be the measure of those rhythms—it could time the Sun or the rate of the pulse. With the clock, time became an objective impersonal dimension. In Lewis Mumford's words, the clock "dissociated time from human events and helped create the belief in an independent world of science".

The clock thus became a regulator of human life. Once clocks became common, the synchronization of activity-at-a-distance, or in complex socio-economic systems, became possible. This gave a vital boost to capitalism. For large businesses needed their workers

"Clocking on." A Dagenham woman checks into the vacuum cleaner factory at which she works. No more dramatic indicator can be found of the transition from earlier times, when the parts of the day were the most accurate time measurements required for practical use: this machine is accurate to the second.

to arrive "on time" (and then to "clock on") and to work "in time". Clock time became the pacemaker for work. Thus the seventeenth-century Puritan Richard Baxter advised his readers: "Do not waste a single moment." Time is scarce. Hence the American advocate of self-help, Benjamin Franklin, advised "do not squander time", for time was becoming valuable: *time was money*. Samuel Pepys took to shaving himself: ". . . it saving me money and time, which pleases me mightily."

The business world of capitalism disciplined its workers to become machines operating according to the laws of clock time, making them work like clockwork, and paying them a rate per hour. And, where efficiency was still lacking, "time-and-motion" studies had to be invoked. Time-keeping became as important as book-keeping. Punctuality became a virtue. Spontaneity yielded to planning, the diary, the timetable—to regularity. In Lawrence Sterne's *Tristram Shandy* the hero's father made love once a month after winding the clock. Similarly, the Enlightenment philosopher John Locke advocated toilet training to produce regular bowel motions. And Lord Chesterfield (1694-1773), writing to his son, aimed to cut down the time thus wasted yet further: "There is nothing which I more wish that you should know and which fewer people do know, than the true sense and value of Time . . . I knew a gentleman, who was so good a manager of his time, that he would not even lose that small portion of it which the calls of nature obliged him to pass in the necessary-house, but gradually went through all the Latin poets in those moments. He bought, for example, a common edition of Horace, of which he tore off gradually a couple of pages, carried them with him to that necessary place, read them first, and then sent them down as a sacrifice to Cloacina; this was so much time fairly gained, and I recommend you to follow his example . . . it will make any book which you shall read in that manner very present in your mind.

"I am sure you have sense enough to know that a right use of your time is having it all to yourself; nay, it is even more, for it is laying it out to immense interest; which in a very few years, will amount to a prodigious capital."

No wonder that anti-urban, anti-bureaucratic, anti-industrial radicals like Rousseau—and modern hippies—began their revolt with the gesture of throwing away their watches.

Other features of modern, complex, urban, industrialized life have reinforced our sense of time as something inexorable, external to us, an objective regulator of existence. Particularly within the Protestant world, the working week became standardized as from Monday to Saturday. The irregular Saint's Day holidays of Medieval Catholicism gave way to the regular Sabbath day of rest. The American critic Marshall MacLuhan has argued that the coming of the printed book—and mass literacy—has reinforced the logic of sequential time. For (unlike, say, paintings) there is a definite order in which a book is read. In novels, in particular, that order is the temporal order of the unfolding of plot and character— ignoring, for the sake of argument, "flashback" techniques and experimental fiction. Hence the overwhelming sense of the grip of history conveyed by the great novelists of the last century—George

Eliot, Stendhal or Balzac and, more recently, Thomas Mann—and Thomas Carlyle's impassioned protest: "O Time-Spirit, how thou hast environed and imprisoned us!"

Subjective Time

Yet, especially amongst writers and artists, the sense of the subjectivity of time has never been lost. Lewis Carroll's Alice found it was always teatime—six o'clock—at the Mad Hatter's Tea Party. In Shakespeare's *As You Like It* Rosalind, meeting Orlando in Arden, engages him in conversation by asking him what o'clock it is. Orlando answers that there is no clock in the forest. Rosalind's reply is that time travels differently for different sorts of people: "Time travels in diverse paces with diverse persons. I'll tell you who Time ambles withal, who Time trots withal, who Time gallops withal, and who he stands still withal." For Macbeth, it was "Tomorrow and tomorrow and tomorrow/Creeps in this petty pace from day to day".

Philosophers likewise have questioned the reality of objective absolute and external Newtonian time. In their different ways the eighteenth-century philosophers Berkeley and Hume both argued that we have no direct experience through our perception of time, but merely of the succession of events. Kant saw time not as a property of the outside world, but as a category of our mind vital to order our experience. Psychologists have investigated how our sense of duration and succession can be grossly distorted by emotions, by the degree of our interest, excitement, boredom or attention. Following this train of thought, the twentieth-century French philosopher Henri Bergson made the distinction between scientific time, and what he called "*la durée réalle*" (lived time) which was "psychical in its nature and psychological in its order".

Other analysts have explored the nature of memory, the faculty which orders our experiences in past time. The mysteries of memory have been debated ever since Plato. Plato suggested that knowledge is not something we pick up from sense experience; it is, rather, innately held information (stored in the soul prior to its experience of the external material world) which can be triggered into recall. Such vagaries of memory led Sigmund Freud to develop the idea that "forgetting" is not a matter of the failure to retain knowledge, but rather an act of "suppression" by our unconscious mental operations, where such memories would be painful or disturbing. In other words, our unconscious continues to be dominated by memories of all experiences.

Furthermore, psychologists have followed up the common claims that some people can experience the future ("precognition"), or have feelings of "*déjà vu*". It has been argued—following J. W. Dunne's *An Experiment with Time* (1938)—that the mind may be able to operate on a number of different time planes, distinct from clock time. Chapter 7 will amplify these points.

Some twentieth-century investigations, of course, mirror the modern revolution in physicists' own notions of time, arising out of the work of Einstein. Einstein's achievement was, in effect, to show that the Newtonian concept of absolute, standard, objective time, uniform throughout the Universe, was meaningless, because no observer could ever actually experience time in that way. Across large distances, simultaneity could not be experienced. This is because the communication of information across the Universe must be *via* waves; such communication is not instantaneous, but takes time. Light (and other electromagnetic) waves travel fastest, but even the speed of light is finite. The further an object is from an observer, and the faster that object may be receding from the observer (approaching the speed of light), the longer its "time" will take to be communicated to the observer. Hence, relative to a (relatively) stationary observer, a clock receding from him at a phenomenally high speed will appear to have a different time pulse from his own; to move slowly, in fact. Einstein's point is that we have no way of escaping from this world of time appearances, because there are no instantaneous connexions between external events and the observer. Time is an aspect of the relationship between observer and Universe, and no observer is in an absolutely privileged position which would make his "time" any more valid than another's.

The physics of relativity, which we shall discuss in more detail in Chapter 5, have reinforced a tendency in modern physical science to break down the regular absolutes of the Newtonian world—the uncertainty principle and quantum mechanics are two other examples. More especially, the consequence of Einstein's physics has been the feeling that time is a rather unsatisfactory concept altogether. Some physicists and mathematicians have tried to ease this by aiming to assimilate it towards the concept of space; in other words, to reinterpret the "time" of an object in terms of the point

(*Above right*) Virginia Woolf, the British novelist whose works explore the nature of subjective time. This same theme is epitomized by the Monet painting (*below right*), "Les Coquelicots".

in space which it is occupying (the dynamization of space). This is the implication of the notion of space time developed by the mathematician Minkowski.

Modern creative writers, however, far from feeling hostile to the new uncertainties of subjective time, have revelled in exploring it. One genre has been the imaginative exploration of relativity itself, and of time travelling, in science fiction, H. G. Wells' *The Time Machine* (1895) being seminal. But practically every experimental movement in modern writing has been preoccupied with the subjective meanings of time and memory, against the backdrop of a society where times change ever faster and "God is dead". Virginia Woolf observed: "Time unfortunately, though it makes animals and vegetables bloom and fade away with amazing punctuality, has no such simple effect upon the mind of man. The mind of man,

James Joyce (1852–1941), the Irish novelist whose works regularly exploited subjective time; his *Ulysses* and especially *Finnegan's Wake* make use of "pauses" in which time passes only in the mind of the "narrator".

A tailpiece by Hogarth: the end of time; his last engraving, executed shortly before his death.

moreover, works with equal strangeness upon the body of time."
Her own novels, such as *To The Lighthouse* (1927), explore personal
time as the sequence of thoughts and experiences connected
together in the individual consciousness by meaningful associations. The expansions and contractions of time in Proust's *A la
Recherche du Temps Perdu* (1922-27) and James Joyce's *Finnegan's
Wake* (1939) similarly explore the unique logic of mental time. The
"new wave" of modern French novelists, such as Alain Robbe-
Grillet and Marguerite Duras, has also been preoccupied with time
as lived experience. They have partly looked back to Bergson's
idea of time as "*la durée réalle*" and have partly borrowed techniques of cinema such as flashforward and flashback (techniques
hardly available to the traditional theatre, even one emancipated
from the Aristotelian unity of time). The Futurist Movement of the
1920s enthused over the modern world's romance of speed. It
glamorized the accelerating pace of change and stigmatized the
old-fashioned. Existentialism has pinpointed once more the absurd

43

and terrifying paradoxes of Man's situation in time, in works ranging from Albert Camus' *The Myth of Sisyphus* (1950) to Samuel Beckett's *Waiting for Godot* (1952). Modern culture has totally dislocated traditional concepts of time.

Coda

It would be presumptuous to believe that we now understand time. Scientific and philosophical investigations produce more new problems than solutions. Yet consciousness of time has changed drastically during the development of human cultures. We now live in a society in which our everyday time experience has become less and less that of natural biological rhythms and conforms ever more to the complex, rational ordering of mechanized work, the city, and the clock. Above all, we live in a society which has changed enormously, is changing ever faster, and whose law is perpetual change (officially interpreted as progress). Hence consciousness of time pervades everything, because time as the agent of change (past and future) dominates our culture. Time has ceased to be cyclical, and has become development. We are all the children of Time. We must not forget that Time devours her own children.

RP

2 The Moving Earth in Space

Since long before the Earth was formed some 4,500 million years ago, its primary star, the Sun, has been pouring its mass into space in the form of energy. The small fraction which impinges on the Earth has enabled the evolution of life itself to take place, and has provided Man with his heat, food and, in more modern times, energy for technological development. For millions of years Man and his ancestors have led lives ordered by the Sun each day and each season. The Sun has fundamentally affected Man's evolution and that of all life around him. It is no wonder that the Sun is Man's oldest god, and is still the most precious manifestation of the modern gods.

To the man of a million years ago, virtually nothing mattered other than the beginning or ending of daylight, or the height the Sun attained in the sky during the day, which affected how warm the man was, and how abundant his food was. This was to provide the foundation of the science of astronomy. Many thousands of years passed before our ancestors studied the behaviour of the Sun, and his pale imitator, the Moon, more closely. But it was realized long ago that the Sun and Moon behaved cyclically. The Sun and Moon would rise in roughly the same direction; they would cross the sky and then, heading for the opposite horizon, set in about the same direction each day.

The great civilizations of the ancient world evolved in moderate latitudes of the Earth where the duration of daylight clearly varied from one-third of the day (a "day" being one complete cycle of the Sun's motion around the sky) in winter to two-thirds in summer.

Ancient astronomers became aware of another movement of the Sun. Its position could be fixed relative to the background stars simply by measuring the angle between the Sun and a bright star in the twilight sky. It became clear that the Sun was in motion across the background of fixed stars, taking one year (a "year" being the time between events such as between one spring equinox

A total eclipse of the Sun.

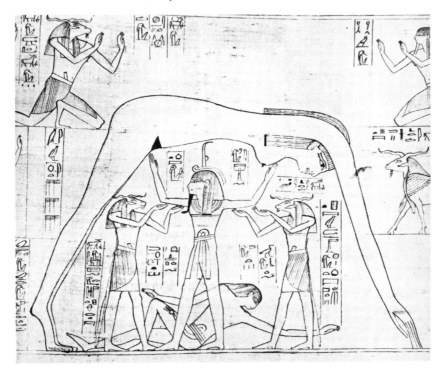

The Egyptian cosmos, one of Man's early attempts to explain the nature of the Universe about him. The goddess Nut, on hands and feet, represents the sky; the god Qeb, reclining below, the Earth; and the god Shu, standing, the air—Shu has a ram-headed god on either side. This representation is from the funerary papyrus of Princess Nesitanebtashru, about 970BC.

Claudius Ptolemy (born *c.*AD75), the Greek astronomer whose geocentric view of the Solar System, with the planets moving on circular cycles and subsidiary circular epicycles around the Earth, held up the progress of astronomy for about a millennium and a half. The reason? His system was better for navigation and timekeeping than early attempts to describe the planets as circling the Sun.

and the next, or between successive midsummer days) to complete one tour of the "celestial sphere", the apparent sphere of fixed stars at infinite distance from us that surrounds the Earth and apparently rotates about it each day. By studying the position of sunrise or sunset relative to fixed alignments of stones and natural objects such as mountain peaks, many ancient tribes of Man were able to observe that the Sun rose increasingly further north (in the northern hemisphere) from the second half of winter up to the middle of summer when sunrise and sunset were furthest north of all. From many years of observing the approximate time of midsummer day, it was seen that there were about 365 days (passages of the Sun across the sky) for one year (passages of the Sun across the celestial sphere of stars). There were also about 12 "moons" corresponding to one cycle of the Moon's phases.

These were the observed data on which the famous Greek astronomer Ptolemy built his picture of the Universe, a picture that was to endure for over one and a quarter millennia, and be

A sixteenth-century representation of the geocentric system. At the centre are the Earth and the other three Aristotelian elements, water, air and fire (in outward progression); outside these are the spheres of the Moon, Mercury, Venus, Sun, Mars, Jupiter, Saturn and the fixed stars.

responsible for such a devastating retardation of the progress of astronomy. Ptolemy's model put the Earth at the centre of the Universe (where, in all fairness, it clearly seemed to be). The Sun, Moon, planets and stars all circled the Earth at their respective distances, held on perfectly transparent rotating crystalline spheres. The stars formed the outermost sphere, and it was against this background that all the other heavenly bodies moved.

Of course this model was clearly too simple to fit the observed motions of the planets and to predict their positions with even elementary accuracy which, being devoted mathematicians, the Greeks wanted to do. Accordingly the model became complicated by systems of epicycles, equants, and deferents that almost need a modern computer to master. But the system did at least explain the observed positions of celestial objects, and it provided an explanation of the variable lengths of the days and the changing seasons.

When Copernicus tentatively suggested putting the Sun instead of the Earth at the middle of the Universe, the explanations became easier in some cases, and harder in others.

A miniature from a thirteenth-century manuscript showing an astronomy lesson in progress; the figure in the centre is using an astrolabe.

Nicolaus Copernicus (1473–1543), the Polish astronomer whose work in establishing a heliocentric model of the Universe at last found a chink in the armour of Ptolemy's geocentric system.

The Day

What is a day? In common parlance it means two things; it means the time during which the sky is bright, as compared with night, when the sky is dark; and it means 24 hours on the clock, during which one date remains on the calendar   In Copernicus' heliocentric model of the Universe, these two types of day correspond to two effects of the rotation of the Earth about its polar axis. The first, the hours of daylight, is simple to understand, and corresponds exactly with the Ptolemaic model. The sky is bright while the part of the Earth you inhabit is in the sunlight. Whether sunset is caused by the Sun disappearing under the Earth, or the Earth turning away from the Sun and taking you with it, is immaterial.

The real "day", the day of 24 hours, is more complicated. In the geocentric (Earth-centred) Universe, you can observe the passage of the Sun across the sky by the shadow cast on a sundial. When the shadow of the gnomon reaches the same point as yesterday, the Sun has made one turn about the Earth and that is one day of 24 hours. But in fact the Earth turns on its axis *and* revolves around the Sun in its yearly orbit, so there is an immediate problem.

50

The only way one rotation of the Earth can be measured precisely is by comparing it with the position of the stars, objects so distant that the yearly motion of the Earth about the centre of its orbit, 150 million kilometres away, is immaterial. Thus a complete rotation of the Earth is the time taken for a star to return to the exact point in the sky where it was previously observed. In other words, one rotation of the Earth in the Copernican system corresponds with one rotation of the sphere of fixed stars in the Ptolemaic Universe.

Apart from their astrological significance, the positions of the fixed stars were not important to the geocentric astronomers, and had no bearing in determining the length of a day. In fact, it would seem quite stupid to use the fixed stars for this purpose: the Sun is moving across the star sphere so that, were we to fix the middle of a day by the culmination of a particular star (the moment the star is highest in the sky when it is exactly due south, as observed from

Copernicus' heliocentric system, as shown in his *De revolutionibis orbium cælestium*.

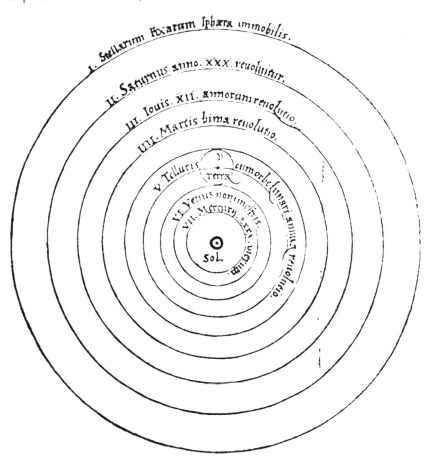

Galileo Galilei (1564–1642), the Italian physicist and astronomer whose feud with the Vatican over the validity of the Copernican system has become part of the mythology of the history of science: although the traditional versions are romantic, it seems almost certain that the Roman Catholic hierarchy had no wish to try him for heresy, and would not have done so had it not been for Galileo's incessant pestering of them; his subsequent recanting is rather less admirable in this light. His trial did, however, do much to publicize the heliocentric ideas.

the northern hemisphere), it might be the Sun was high in the sky—yet six months later we would be in darkness at "midday" as determined by the star.

The reason for this is that the Earth turns about 365 times on its axis as it makes one journey around the Sun; in one day the Earth travels about 1° of its 360° orbit round the Sun. From Earth, this makes the Sun appear to move 1° eastwards each day *relative to the stars*. Conversely, any particular star will be 1° further west at the same sundial time each day. So the sidereal day, the day measured by the stars and corresponding to one rotation of the Earth on its axis, is not used for the basis of our clock time. The extra 1° gained by the Sun each day means that the Earth takes an additional 1/365 of 24 hours to complete one revolution relative to the Sun; 1/365 of 24 hours is just under 4 minutes.

Sidereal time is of no significance in the daily lives of most of the creatures on Earth. Astronomers are the exception; they need to know the position of stars in the sky relative to their telescopes, and must keep the telescopes pointing accurately at the star while the Earth rotates relative to the celestial sphere of fixed stars. Accordingly, observatories have sidereal clocks which are regulated to gain one day exactly in one year, just under 4 minutes per day, and the telescopes are driven at one revolution per sidereal day to follow the stars.

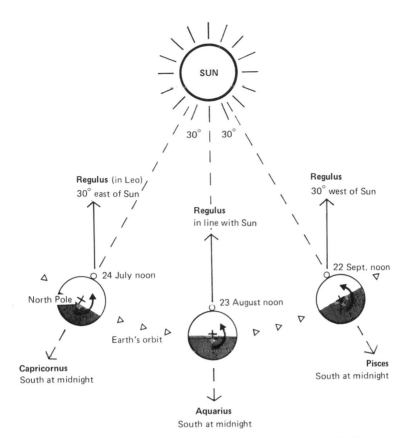

An observer in the northern hemisphere observing the positions of the Sun and
the star Regulus in the constellation Leo on three dates, approximately one month
apart, during the summer. Three different zodiacal constellations are due south at
midnight on each of these dates. Due to the apparent easterly motion of the Sun
among the stars, the Earth turns once on its axis *relative to the stars* in about four
minutes less than the full twenty-four hours. (Not to scale.)

The Month

If you are looking through a telescope at the Moon, (or the Sun in
theory at least: looking at the Sun through a telescope is a certain
way of blinding yourself), then the telescope drive must be regulated to allow for the apparent movement of the Moon or Sun across
the celestial sphere. The apparent motion of the Moon is faster
than that of the Sun. Starting from new moon one month, when the
Moon is aligned in the sky with the Sun (although not necessarily
in front of the Sun's disc, which happens only during an eclipse of
the Sun), the Moon makes one tour of the celestial sphere eastwards in just over 27 days. This is called its sidereal period and is
the true time the Moon takes to make one orbit of the Earth. But in
one sidereal month the Sun has moved eastwards on the celestial
sphere by about 27° which the Moon takes over two days to cover,

when new moon occurs once more. The total time taken for the Moon to go from new to new, overtaking the Sun at about 12° per day, is called the synodic period, and is 29½ days. There are just over 12 synodic months in a year, corresponding to the "moons" of the Red Indian and other races, the most natural month (the word "month" also means "moon") of the several possible definitions.

However, the motions of planets and satellites around their primary bodies are not uniform: because of variations in the motion of the Moon around the Earth and of the Earth/Moon system around the Sun, the lengths of sidereal and synodic months also vary, and the values given above are averages. This further complicates the definition of time by the Sun, which is the basis of our daily time-measuring system.

Time by the Sun

Before the invention of mechanical devices for indicating the passage of time with any accuracy, the closest anyone needed to know the absolute time, as opposed to an interval of time, was the time dictated by the Sun. Until very late in our history, sunrise, noon and sunset were sufficient times for most people, indicating when you began work, when you ate, and when you went to bed. An obvious refinement of this was the sundial, and, as we shall see in the next chapter, many different designs exist. It was not long after the invention of the sundial that, by comparing it with other methods of time measurement, it became clear that the Sun could be "fast" or "slow" depending upon the time of year.

At noon by the clock, in November the Sun is found to be almost 17 minutes past the meridian, the imaginary line passing across our sky from the zenith (overhead) to the south point on the horizon. ("Meridian" means "middle of the day".) In February the Sun is almost 15 minutes slow, whereas during the spring and summer months it gains and loses between four and six minutes in two cycles. In fact the sundial is accurate on only four days of the year, about April 15, June 14, September 2 and December 25 (see page 64).

This apparent non-uniform motion of the Sun is, as we have mentioned, due to the non-uniform motion of the Earth in its orbit, and it is combined with another effect, due to the inclination of the Earth's axis, to which we will return shortly.

After Nicolaus Copernicus (1473-1543) freed men's minds from the shackles of geocentricity, which had been so securely fastened by Ptolemy (although, four hundred years before Ptolemy, the

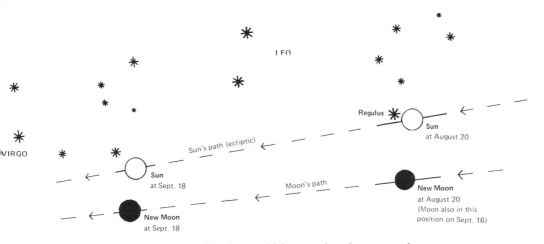

Comparison of the positions of the Sun and Moon against the stars on dates one synodic month apart. (Note that the Moon's positions for these months in the following years would be different, because the position of the Moon's path across the stars changes noticeably with time; also that the disks of the Sun and Moon are not to scale with the background.)

Greek Aristarchus had suggested that the planets circle the Sun), the planetary motions became easier to understand in a basic sense, but precision measurement and prediction of their positions became more difficult. In fact Copernicus and many of his contemporaries reverted to epicycles superimposed on the circular planetary orbits to explain why the planets, including the Earth, were sometimes ahead of and sometimes behind their predicted positions, with a variation that had all the characteristics of a circular function such as an epicycle.

One problem was the lack of really accurate positional data. The telescope had yet to be invented: all measurements had to be made with instruments using the naked eye alone. Fortunately for astronomy two men whose interests in the subject were quite different were thrown together at this time: Tycho Brahe (1546–1601) and Johannes Kepler (1571–1630). Tycho Brahe, a Dane of noble descent, during his colourful lifetime set up what must have been the most magnificent naked-eye observatory ever, on the island of Hven off the Baltic coast of Denmark. This island, a present from King Frederick II of Denmark, was ruled by Tycho in what can be described at best as a rather tyrannical fashion. He built his observatory, called Uraniborg, and equipped it with instruments of unequalled accuracy; for twenty years he collected the most precise and comprehensive measurements of the stars and planets that had ever been made. In 1596, following the death of his royal patron,

(*Above*) Tycho Brahe's mural quadrant showing himself and the instruments he used at Uraniborg; a picture from his *Astronomiae instauratae mechanica*.
(*Below*) Brahe's planetary system showing the Sun, around which travel the planets and the comet of 1577, itself travelling around the Earth and Moon; from his *De mundo aetherei recentioribus mensis*.

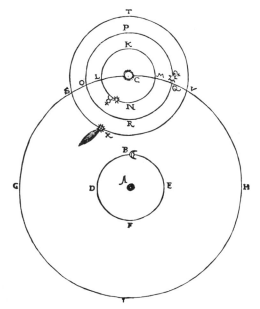

IOANNIS KEPPLERI,
Mathematici Cæfarei
hanc Imaginem,

ARGENTORATENSI BIBLIOTHECÆ
Confecr.

Johann Kepler (1571–1630), the German astronomer, a pupil of Brahe, whose brilliant work on planetary orbits at last put the final nail into the coffin of the Ptolemaic system while at the same time providing a platform for Newton to base his famous laws of motion on.

Tycho left Denmark because of his worsening relationship with the people of Hven and the new members of court. Eventually he set up an observatory in Prague where he was joined by a young man whose first book had impressed the ageing Tycho. The young man was Johannes Kepler.

It was Kepler who, using Tycho's measurements, collected with such persistence over the twenty years at Uraniborg, made the final step that took astronomy out of the dark ages. Tycho himself was a firm believer in an Earth-centred Universe, although he devised a hybrid system of his own in which the Sun circled the Earth and certain of the planets circled the Sun. Kepler was deeply interested in astrology. But, together with Copernicus, these two men were finally to divorce astronomy from astrology and pave the way for the work of Newton and Einstein.

The breakthrough Kepler made was to deduce empirically from Tycho's observations, particularly those of the planet Mars, that the planets did not move in circles and epicycles, but in elliptical orbits with the Sun at one focus of the ellipse. Thus the orbits were slightly eccentric: the planets were closer to the Sun at one part of their orbit than at the opposite point. Moreover, the motion of the planet around its elliptical orbit was not uniform; the planet travelled faster the nearer it approached the Sun, so reaching a maximum angular speed around the Sun (and maximum velocity through space) at the closest point, perihelion.

The Earth reaches perihelion in early January, just after the winter solstice (the "shortest day") in the northern hemisphere. It is then at a distance of 147 million kilometres from the Sun. At the opposite point on its orbit, aphelion, which the Earth reaches early in July, it is 152 million kilometres from the Sun.

At perihelion the apparent motion of the Sun across the celestial sphere—that is, the daily motion of the Earth around its orbit—is a little over 3% faster than it should be if it were to keep perfect sundial time, so each successive day at about this time of year the Sun will be seen due south a little bit later than it should. The error starts increasing as winter sets in and diminishes as spring approaches so that, starting from its 17-minutes-fast position, towards the beginning of November, the Sun begins to lose time at an increasing rate until perihelion at the beginning of January, by which date the sundials are already slow and getting slower, until mid-February when the Sun appears to speed up once more.

This yearly cycle causes a variation in apparent (also called "true") solar time, as read on the sundial, which variation has a

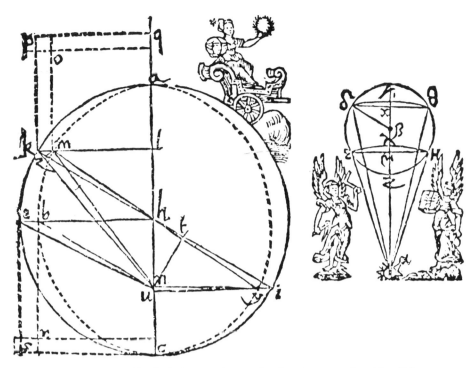

An illustration from Kepler's *Astronomia nova* showing the eccentric orbit of Mars; it was in this book that he announced his preliminary results on the elliptical, as opposed to circular, heliocentric orbits of the planets.

frequency of one year. In other words, were this the only effect on apparent solar time, sundials would be fast in the autumn and slow in the spring and that would be all. Unfortunately for solar time, however, there is another effect to be taken into account.

The Seasons

The Sun moves across the celestial sphere of fixed stars once each year in a sharply defined path. This path can be—and is—drawn on star maps: it is called the ecliptic, and represents the plane of the Earth's orbit extended infinitely until it reaches the celestial sphere, which as we have seen, is assumed to be a sphere of infinite radius. The ecliptic marks the yearly motion of the Sun; either side of it is a fairly narrow band of sky in which all of the planets (except Pluto), many of the asteroids, and many other of the bodies in the Solar System are to be found, because most of the Sun's family have orbits in much the same plane in space. This band of sky takes in the star groupings (constellations) Aries, Taurus, Gemini, Cancer, Leo, Virgo, Libra, and so on; that is, the familiar constellations of the Zodiac.

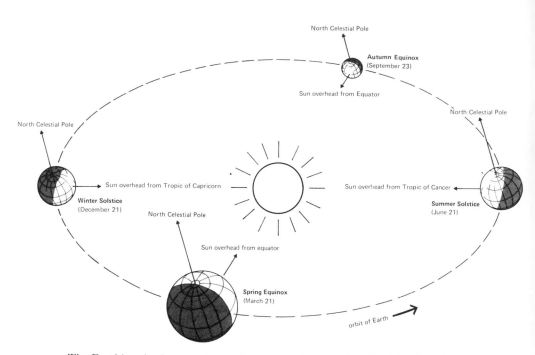

The Earth's axis always points to the same position on the celestial sphere during the year—i.e., its annual journey around the Sun—giving rise to the seasons. Here we see the Earth at the two solstices and two equinoxes. (Not to scale.)

The Earth rotates daily on an axis which is roughly at right angles to the plane of its orbit, the ecliptic. It is only very roughly at right angles; in fact there is a 23½° angle of inclination relative to the true right angle—in other words, there is a 23½° angle between the plane of the ecliptic and the plane of the Earth's equator. The Earth spinning on its axis has a considerable angular momentum, and, since a considerable force would be necessary to change the direction of this momentum, the spinning Earth acts like a gyroscope and its axis, to all intents and purposes, is always oriented in space in the same direction. In effect, we can extend the line of the Earth's axis from the poles into space until it meets the celestial sphere; the two points where it does so are called the celestial poles. All stars appear to revolve about the celestial poles each day due to the Earth's axial rotation, but the celestial poles are fixed. There is a fairly bright star called Polaris near the North Celestial Pole. Although this star does make a tiny circle in the sky, because it is about 1° from the true North Celestial Pole, it is near enough to act as a very good pole star.

If the Earth's axis were at right angles to its orbit—that is, aligned with the axis of the ecliptic—the ecliptic would be exactly overhead for someone standing on the Earth's equator; the Sun would rise due east, cross directly overhead at the equator and set due west. Each day would be of the same length, no matter where on Earth you were. At the poles, the Sun would be cut in half by the horizon as it made a tour right round it.

In reality, this is the case only at the equinoxes, when the Sun *is* overhead at the equator. Due to the inclination of the polar axis, the Earth's northern hemisphere is inclined away from the Sun during the months from mid-September to mid-March. At the same time, the southern hemisphere is inclined towards the Sun. The reverse is true during the other months. Thus, the Sun is higher in the sky on successive noons as the midsummer solstice approaches, and thereafter it is lower at noon until the midwinter solstice.

The effects are complex, and far-reaching. Due to the increasingly direct sunlight and heat, the summer months are much hotter than the winter months in the temperate regions of the Earth. Near the equator, where the Sun is never far from being overhead at noon ($23\frac{1}{2}°$ is the most it can be, at either of the solstices) the climate is tropical and, apart from the rainy seasons, varies little. But in intermediate latitudes the result of the Earth's inclination is the changing seasons, with the obvious effects these have on the lives of the animals and plants which inhabit the planet. The further from the equator you are, the greater is the contrast between summer and winter.

At noon on midsummer day in the northern hemisphere, for example, which occurs on about June 21, if you could travel instantaneously north from the equator you would reach a point where the Sun was overhead. This would be at a latitude of $23\frac{1}{2}°$ North, the Tropic of Cancer, which marks the most northerly point where the Sun can ever be directly overhead.

Further north on the summer solstice noon, the Sun is now on the meridian, where we have temporarily halted it, in the southern part of the sky. The celestial equator overhead at the beginning of our journey, crosses the meridian to the south, but lower and lower in the sky as we speed further north. We reach a latitude where the celestial equator crosses the meridian due south at an altitude of $23\frac{1}{2}°$. In the opposite direction, due north, the celestial equator is $23\frac{1}{2}°$ below the horizon: were we to rest on our journey at this point and allow twelve hours to pass, we would find that at midnight the Sun was due north and just on the horizon. It must be so, of course, since the Sun at this moment is $23\frac{1}{2}°$ north of the

The midnight Sun: at midsummer on the Arctic or Antarctic Circles, the Sun remains above the horizon throughout 24 hours. The closer to either pole, the longer the midsummer "day".

celestial equator. We have reached the land of the midnight Sun. In a few days time, the Sun will have continued its journey round the ecliptic and moved a degree or so south, and will now just disappear at midnight. If we then took a much longer break and waited there for six months, we would find that, on the day of the midwinter solstice, the Sun did not rise, even at noon. The place at which we have rested so long is at latitude 66½° North, 23½° from the North Pole, and is on the Arctic Circle.

Finally we must spend a year at the North Pole itself. On this midsummer day we can watch the Sun, at an altitude of 23½° in the sky, make one complete turn around the sky parallel to the horizon. We have some difficulty deciding when it is noon, because all directions are south from here! Each great circle of longitude passes beneath our feet, so we can say that when the Sun passes over 0° longitude it is true solar noon on the Greenwich meridian.

As the days wear on, the Sun continues to make its complete tour of the sky parallel to the horizon, but getting a little lower each day until, at the autumn equinox, the Sun is bisected by the horizon. Within a couple of days it has gone, and will not be seen again for almost six months. Thus at the poles half the year is perpetual day and the other half perpetual night.

In fact, due to the acceleration of the Earth past perihelion, the Sun is south of the equator for several days less than it is north of the equator. Which brings us back to the Equation of Time, the curve of Sun-fast Sun-slow that must be applied to sundial readings, unless the correction is built into the sundial itself.

The Equation of Time

The ecliptic is a great circle around the celestial sphere which crosses the celestial equator at $23\frac{1}{2}°$ at the points the Sun occupies at the time of the two equinoxes. Astronomers conventionally have used the position of the spring equinox, called the first point of Aries (although, as we shall see, it is now in fact in the constellation Pisces) as a point from which to make measurements. It is the Greenwich of the skies.

The time indicated on the sundial by the Sun in its daily progress across the sky depends upon its westerly progress alone. Its height above the equator (called *declination*, which is a celestial equivalent of latitude) has little effect on the direction of the shadow on the sundial plate. The direction of an object—east, south-west, etc.—is its *azimuth*. It is the variation in the Sun's azimuth from its mean daily position which determines the accuracy obtainable using a sundial.

The Sun travels just under 1° eastwards along the ecliptic each day. In March, near the equinox, when the Sun is climbing rapidly northwards in the sky, the daylight hours are noticeably lengthening in the northern hemisphere. When the Sun is near the summer solstice (or winter solstice, for that matter) its 1° of easterly motion makes little difference to the length of the day. Sunrise and sunset times are dependent only upon the Sun's distance north or south of the equator. At the equinoxes, one day's motion along the ecliptic of about 1° represents only about 0.9° progress eastwards, and 0·4° north or south. At the solstices, the 1° is almost entirely in an easterly direction, and there is virtually no change in the declination, the north-south position of the Sun in the sky, nor, consequently, in the sunset and sunrise times.

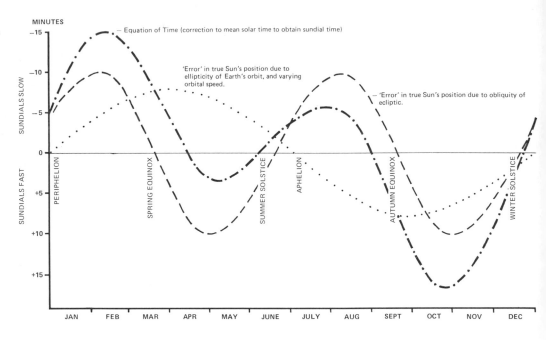

MINUTES

— Equation of Time (correction to mean solar time to obtain sundial time)

'Error' in true Sun's position due to
ellipticity of Earth's orbit, and varying
orbital speed.

— 'Error' in true Sun's position due to
obliquity of ecliptic.

SUNDIALS SLOW

SUNDIALS FAST

PERIHELION • SPRING EQUINOX • SUMMER SOLSTICE • APHELION • AUTUMN EQUINOX • WINTER SOLSTICE

JAN   FEB   MAR   APR   MAY   JUNE   JULY   AUG   SEPT   OCT   NOV   DEC

Curves showing the time gained and lost by the Sun compared with the uniform
motion around the celestial equator of the fictitious 'Mean Sun', due to: (1) the
non-uniform motion of the Earth (the Earth's orbit is an ellipse, not a circle); and
(2) the obliquity of the ecliptic (see text). When these two factors are added
together the curve of the "equation of time" results; this is the correction that
must be made to mean time to give apparent (i.e., sundial, or true) time.

But, as far as the sundial time is concerned, the Sun's rapid
easterly progress at the solstices compared with the equinoxes
means that the Sun is gaining at both of the solstices and losing at
both of the equinoxes. This gives a cycle of variation in apparent
solar time which has twice the frequency of the main variation due
to the motion of the Earth around its elliptical orbit. Because of the
fairly close coincidence of the solstices with perihelion and aphelion, the two effects tend to add together in February and
November when the biggest sundial errors occur.

Long before the mechanical clock became a common instrument
of time measurement, the Sun's deviation from mean time was
known. But this deviation, which was known to the Greeks, was
not large enough to be troublesome in daily life. In more recent
ages, precise timing has become increasingly important. Accordingly, clock time is based on a fictional "mean Sun" which is
assumed to move at a uniform rate around the celestial equator,
rather than the ecliptic, corresponding to the 0·98651° per day
average daily motion of the true Sun. This is the basis of mean
time, which is used throughout the world today.

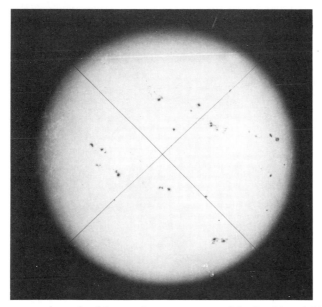

One of the more enigmatic of the regular occurrences of the Earth's neighbour hood is the sunspot cycle, which is to us an extremely important one since it affects our upper atmosphere and our radio communications. Approximately every eleven years the number of spots on the Sun's disk reaches a maximum, as above. (*Below*) A "close-up" of a sunspot.

With the improvement of world-wide communications, both in travel and with the advent of radio and telephonic media, the concept of mean time, as applied locally, also became inadequate. When it is noon Greenwich Mean Time, the "mean Sun" is due south, on the meridian, at Greenwich Royal Observatory (near London). It is only just rising in the eastern states of the United States, and has set in Tokyo. Just as it would be inconvenient to use sidereal time, so it would be inconvenient for all the world to use the same mean time. On the other hand, every place on Earth with a different longitude has a different local mean solar time, as well as different local apparent solar time and local sidereal time!

The local sidereal time is important if you are an astronomer and wish to point your telescope at an object of known position in the sky. When you are on the latitude of Greenwich, but to the west of Greenwich by 69 kilometres, you are 1° of longitude west; therefore an object due south exactly at noon as seen from Greenwich will not be due south from your observatory for another 4 minutes.

To find the true local time, therefore, the mean time must be adjusted according to your longitude. Someone at a place 15° west of Greenwich will not see the object due south until one hour later than if he were at Greenwich. So if the date is April 15 when the Equation of Time (the correction to find true solar time) is zero, at Greenwich the Sun will really be on the meridian at noon GMT. But in Cork, in the Republic of Ireland, on a clock reading GMT, the Sun will not be due south for over half an hour. In New York, 74° west of Greenwich, the Sun will not be on the meridian for almost five hours. Each 15° of longitude represents an hour's difference in true local time.

Since 1880, when Greenwich Mean Time was adopted in Britain as the legal definition of 'time', other countries have gradually adopted GMT as the basis for their own time, correcting their local time by one hour for each 15° of longitude east or west of Greenwich. In theory, the world is therefore divided into 24 time zones, with their centres at 15° intervals from each other and thus each differing by one hour from the adjacent zone. Within one zone, the true local mean time can differ from the mean time in the zone by up to half an hour, plus or minus. In practice, the time zones do not follow the longitude divisions as precisely as that, but it is close enough to estimate mean time in most places in the world merely by knowing the longitude.

In North America, the width of the continent is such that five mean time zones are required. Atlantic Time, four hours behind

Halley's Comet. The English astronomer Edmund Halley (1656–1742), while working on his observations of the comet of 1682, realized that its path resembled those of the comets of 1456, 1531 and 1607 and suggested that all four were in fact a single comet in orbit around the Sun with a period of about 75 years; he therefore predicted that it might return again in about 1758. Although he did not live to see it, the comet's reappearance at the end of the year and the consequent success of his prediction provided a dramatic confirmation of Newton's theories and showed yet another regular cycle of the cosmos.

Greenwich, applies in Greenland, Labrador, and the east coast of Canada. Eastern Time, five hours behind GMT, includes New York and the eastern United States, Quebec, the West Indies, and so on. The other time zones, reading westwards, are Central, Mountain, and Pacific, the latter being eight hours behind GMT and applying in the west coast states of Canada and the USA.

In any of these zones, the local time can differ by half an hour from the official zone mean time. So setting up a sundial requires correction for the Equation of Time, and for the longitude of the situation. The sundial can be set permanently to correct for its longitude, since the error is constant, but for the Equation of Time either a correction must be applied to each reading or an anelemmatic sundial must be used. This has a specially designed gnomon, or shadow-casting device, and the position of the shadow must be interpreted according to the date.

Astronomers have, in more recent times, needed to measure time more accurately even than mean solar time. At first, GMT became known as Universal Time, UT, starting at midnight with 00.00 hours, and counting the following 24 hours. This was regarded as

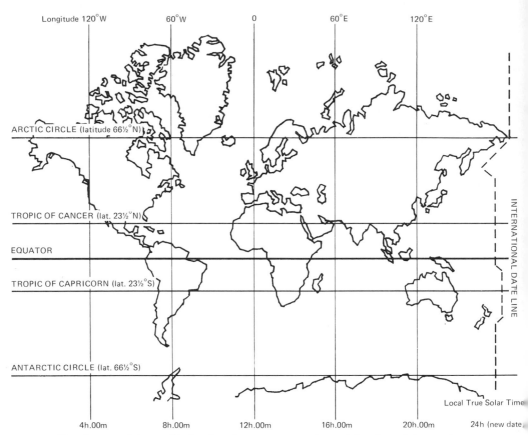

Longitude 120°W   60°W   0   60°E   120°E

ARCTIC CIRCLE (latitude 66½°N)

TROPIC OF CANCER (lat. 23½°N)

EQUATOR

TROPIC OF CAPRICORN (lat. 23½°S)

ANTARCTIC CIRCLE (lat. 66½°S)

INTERNATIONAL DATE LINE

Local True Solar Time

4h.00m   8h.00m   12h.00m   16h.00m   20h.00m   24h (new date)

Sketch map of the world showing the equator, tropics and Arctic and Antarctic circles, together with some longitudes and their corresponding local mean times at noon GMT.

an improvement on the original convention of measuring GMT from noon rather than midnight. UT was precisely related to sidereal time, the observed rotation period of the Earth, but it was then discovered that the rotation of the Earth was not precise enough: it was subject to a number of regular and some irregular variations due to tidal and dynamic effects that cause both long and short-period variations. Sidereal time as observed by the transit of certain specified stars was therefore not sufficiently accurate for the timing of events remote from the Earth. The irregularities in the Earth's motion recalled the ghosts that had haunted the Greeks, and later Copernicus and Kepler.

Over two centuries ago it was speculated that the rate of rotation of the Earth might be affected by the friction of tidal forces raised by the Sun and Moon. As the motion of the Moon was more closely studied and measured, it became evident that its position did not precisely coincide with its ephemeris, the tables of calculated positions worked out from Newton's Laws of Motion. This provided clear evidence that the rotation of Earth, and hence the basis of time measurement, was not uniform. In fact, discrepancies in the ephemerides for all the planets were also present due to the same effect, but, because of the Moon's relatively rapid daily motion as seen from Earth, its very small errors were easier to detect.

One effect, recognized in the middle of the last century, is termed secular retardation. Theory predicted a very small acceleration of the Moon from its expected position due to the effects of gravitational perturbations, but in fact it was found that this acceleration, which would be evident only over long periods of time (hence "secular"), was less than expected. This reduction in the expected acceleration, the secular retardation, it was concluded, was due to variation in the Earth's rate of rotation. Since then a yearly variation and various irregular variations have all been discovered. The most important effect of all these is that due to the tidal forces of the Moon. The gravitational forces attracting the Earth and Moon towards their common centre of gravity, beneath the Earth's surface, cause tides to be raised in the solid surfaces of both bodies and, most noticeably, in the oceans of the Earth.

To the mariner, the state of the tide is almost the only clock he needs when in port and, once again, the clock is an astronomical one. Basically, there are two tides in just over a day, due to the raising of two 'bulges' in the oceans at positions opposite each other on the Earth. The bulges are due mainly to the gravitational attraction of the Moon (the Sun also has a small effect). Since the Moon's attraction diminishes as the distance from the Moon increases, in proportion to the square of that distance, water nearest the Moon is attracted more than the Earth itself which is, in turn, attracted more than the water on the side of the Earth furthest from the Moon. Of course, since the Earth's mass greatly exceeds that of the Moon, the acceleration towards the Moon is only small, but the accelerating *force* is there just the same. The bulges do not "line up" with the Moon but lag behind due to friction; the result of this is that the Earth itself suffers a very gradual slowing of its rotation.

At the same time, the Moon also is subject to tidal forces, due to the presence of the Earth. These have long since reduced the Moon's axial rotation until it is synchronized with the period of revolution around the Earth, so that the Moon keeps one face turned to the Earth throughout the month (except for a slight

Two of the instruments of the Royal Observatory at Greenwich. (*Above*) The great equatorial telescope, from a late nineteenth-century work, *Old and New London* by Edward Walford. (*Below*) Airy's transit circle.

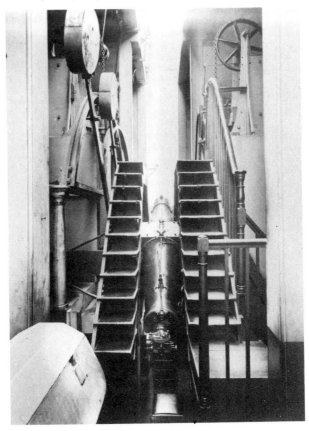

apparent "wobble" due largely to the ellipticity of its orbit and variation in its speed as the Moon moves around it). Since the laws of conservation of angular momentum demand that these tidal forces must be directed somewhere, the result is the very gradual secular acceleration of the Moon.

The variations in the Earth's rotation, when averaged out, represent a slowing down. But the shorter-period fluctuations can reduce the effect or temporarily cause it to disappear. The net result is that the Earth will "lose" about a day in 7,500 years.

With the development of instruments such as the caesium clock, which depends for its operation on the incredibly precise rate of decay of a caesium isotope, the measurement of time has become independent of astronomical phenomena. Man-made clocks are now used to time events in astronomy, although we can never escape from the fundamental cycles of the Sun's passages across the sky and the seasonal variations which we discussed at the beginning of this chapter.

Precession of the Equinoxes, and Similar Matters

As well as the times of sunrise and sunset, their variation through the year, and the differing positions of the Sun in the sky, those who are familiar with the stars associate the progress of the year with the appearances and positions of different stars and constellations in the sky. As the Sun moves across the celestial sphere, the portions of the sky more remote from its glare become visible. Thus on midsummer day (the middle of winter in the southern hemisphere) the Sun is as far north as it can get, just entering the constellation of Gemini. The magnificent constellation of Orion is just below the Sun in the northern skies and is therefore quite invisible. Six months later, however, the Sun is diametrically opposite Orion in the sky, near the constellation of Scorpio. So to northern observers of the stars, Orion is associated with winter and Scorpio with summer, while to southern observers, the position is, of course, reversed. As the northern autumn wears on there is always the unique occasion when, being out exceptionally late, the sky watcher suddenly sees Orion shouldering his way over the horizon almost exactly in the east. Since Orion tends to disappear sometime late in March the return of this familiar friend is always an occasion to remember even if it brings a shudder at the prospect of the winter signalled by his return.

Other constellations are also associated with times of the year: Leo with the spring, Scorpio with the summer, the Great Square of

A representation of the zodiac from Roman times, dating from the first century AD.

Pegasus with the coming of autumn, and so on. To southern observers, of course, the seasonal associations of these constellations is the opposite to those of observers in the northern hemisphere.

Ancient astrologers marked the positions of the Sun, Moon, and planets in the sky against the background of fixed stars within the narrow band of the zodiac. This was divided into 12 equal sections measuring 30° across the sky, corresponding to the distance travelled by the Sun across the celestial sphere in one month (rounding off all the figures, of course).

At the astrologically significant time of the spring equinox, the Sun was in the "House" of Aries: that is, the celestial equator crossed the ecliptic in the constellation of Aries. People born while the Sun was in that constellation were described as being born "under the sign of Aries", and it corresponded to being born between the date of the spring equinox, March 21, and one month later, April 20. The sequence of constellations in the zodiac continued throughout the year in the easterly sequence in the sky: Taurus, Gemini, Cancer, Leo, Virgo, Libra, Scorpio, Sagittarius, Capricorn, Aquarius and Pisces.

The actual point of the spring equinox in the sky, the position of the Sun entering the House of Aries, was called the "First Point of

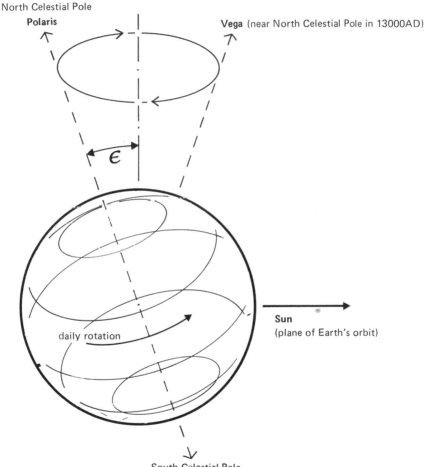

The Earth at the northern winter solstice, showing how the axis of rotation precesses over thousands of years and its effect on the nearness, or otherwise, of bright stars to mark the celestial poles in our skies—as Polaris marks the north celestial pole at present. The diagram shows also that the Sun is invisible all day from the Arctic Circle, and all points north, on this date, even at noon; from points south of the Antarctic Circle the Sun is visible even at midnight. The Sun is overhead on the Tropic of Capricorn. ε is the angle of inclination of the Earth's axis to the line at right angles to the plane of the Earth's orbit; this angle varies over long periods.

Aries". However, the mean Sun is at the "First Point of Pisces" (to coin a phrase) at the solstice, whereas according to the zodiacal sequence given it should be in Aries, one sign to the east. The reason is that the whole sky is subject to another very long-term variation, the precession of the equinoxes.

The plane of the Earth's equator is tilted to the plane of the ecliptic by an angle of about 23½°; that is, the celestial poles are some 23½° from the poles of the ecliptic on the celestial sphere. The celestial poles are virtually fixed and, as described earlier, they give rise to the seasons on Earth. But, as the Earth is not a perfect sphere with an evenly distributed mass, the effect on it of the gravitational forces of other bodies in the Solar System is the same as an out-of-balance force acting on a gyroscope. Just as the gyroscope axis is basically aligned in space, so is the Earth's. But apply an out-of-balance force to the gyroscope and its axis starts to wander through a small circle. Similarly, the direction of the Earth's axis is wandering about the direction of the ecliptic poles in an oscillation which takes about 26,000 years to complete one cycle. As a result, the celestial poles are making circles in the sky about the ecliptic poles; these circles are, in angular measure, each about 23½° in radius. The North Celestial Pole is at present moving towards the Pole Star, Polaris, which it will be closest to in about 150 years' time (it will never actually pass through Polaris). By the year AD10,000, the North Celestial Pole will be fairly close to Deneb, the bright star in Cygnus, and by about AD13,000 the pole will be close to Vega. By AD28,000, the pole will be once more near its present position.

The result of precession is that all the constellations at distances from the pole greater than 23½° must "gain" one sidereal year in 26,000 years. In other words, the stars are rotating relative to the first point of Aries, the spring equinox.

Nowadays, the Sun is in the constellation of Pisces at the spring equinox and, while this point is still sometimes called the first point of Aries, the Sun has moved almost one complete sign of the zodiac, or 30° west of the constellation Aries due to precession since the ancient astrologers named this point. In fact, the Sun is now close to the constellation of Aquarius at the spring equinox. The old astrological "houses" are still used by those who profess their belief in astrology, but these now differ from the astronomical constellations by almost one complete sign of the zodiac.

Not even the angle at which the Earth's axis is inclined to the vertical of the ecliptic is constant. It oscillates between 24° 36′ and 21° 59′ at the same time as it precesses, so that, if the path of the celestial pole through the stars were plotted for 26,000 years or so, it would resemble a gear wheel with rounded teeth, except that it would not close upon itself exactly after one sequence of precession. At present, the obliquity of the ecliptic is decreasing, and it will continue to do so for many centuries as its rate of decrease is only about half a second of arc per year.

Another small disturbance also occurs on the Earth's axial orientation. The plane of the Moon's orbit is inclined to that of the ecliptic by about 5°. This means that at some times the Moon can be up to 5° higher or lower in the sky than the Sun would be when at its closest to that position. For example, at the summer solstice the Sun at noon from London reaches an altitude of 62½° (at that time, as we have already seen, the Sun is just entering the constellation of Gemini). The Moon can be in the same position along the ecliptic, but reach an altitude of 5° higher or lower than that of the Sun; i.e., 67½° or only 57½° .

The Moon's orbit crosses the ecliptic in two places, called nodes. These nodes are not fixed in the Moon's orbit, but move slowly westwards at a rate of 19° per year. In just over 18½ years, therefore, the nodes return to their original positions. So, if the Moon were on the ecliptic at the first point of Aries at the spring equinox, there would be an eclipse of the Sun. Just over 18½ years later the node would be in that position again—only this time the Moon itself would still have over a quarter of its orbit to cover to reach the node, and so there would be no eclipse on that day.

This rotation of the plane of the Moon's orbit, and hence the position of the Moon relative to the Earth through the month, also exerts on the Earth a gravitational force which has a cycle of just over 18½ years; this is called nutation. The Earth's axis nods back and forth over a few seconds of arc.

As a result of all these disturbances, the observed sidereal time, determined by the culmination of any particular star, is not absolutely uniform. Accordingly, in the 1950s the International Astronomical Union adopted Ephemeris Time as the standard for astronomy. This is based entirely on the ephemerides of the Moon, Sun and planets, and so is independent of the Earth's motion and is fixed by observation. The measure of Ephemeris Time was chosen to agree as closely as possible with Universal Time measured during the last century and started from 12 hours GMT, 0 January 1900. The Sun's position at that instant became the

The phases of the Moon. As the Moon travels around the Earth (which itself is travelling around the Sun), different proportions of the lunar disk, as seen from the Earth, are illuminated by light from the Sun. Here we see the Moon in its first quarter (*above left*), seven days after New Moon; fifteen days old, when it is more or less full (*above right*); aged twenty days, when it is approaching its third quarter (*below left*); and aged twenty-seven days, showing just a faint crescent a day or two before the subsequent New Moon (*below right*).

point at which the *tropical year* could then be defined: the tropical year was the time for the Sun to return to this exact position as measured for the equinox of date. In other words, effects such as obliquity and precession are taken into account automatically in the definition of the interval of the year, so that the Ephemeris Year is independent of any variations whether irregular, periodic, or secular. The Ephemeris second is then merely a stated fraction of this year: there are 31 556 925·974 7 seconds in the tropical year.

By 1978, the difference between Ephemeris Time and Universal Time amounted to about 49 seconds, and the most the difference is likely to be this century is a few minutes. The irregularities are far too small to be noticed in everyday phenomena, such as sundial readings, so the difference is important only to astronomers.

Eclipses and the Saros

So much for the effect of the daily and yearly motion of the Earth on the development of time measurement. The day, month and year are entirely functions of the motion of the Earth/Moon system. But there is one other time function—which was, quite incredibly, known to some of the most ancient of astronomers, in the Egypt of the Pharaohs.

As we have seen, the plane of the Moon's orbit is inclined to that of the Earth's orbit by just over 5°. The positions where the Moon's orbit crosses the ecliptic are called nodes and the Moon must pass through two nodes each month. The nodes move slowly westwards along the ecliptic at a rate of some 1½° each month; so, to all intents and purposes, the nodes of the Moon's orbit do not change much in position over a year. The Sun, in its easterly journey along the ecliptic, must pass through these two nodes each year at an interval of about six months. Since the Sun moves only 1° each day and the Moon takes only 15 days to pass through half its synodic period—that is, to revolve about the Earth from one alignment of the Earth and Sun to the next—the Sun cannot be more than about 7° from the node (and hence the Moon more than 7° from the same or opposite node) at one of these alignments. (New Moon or Full Moon can be referred to as "conjunction" or "opposition", which cover all close alignments of the Sun, Moon, planets and other bodies; or the term "syzygy", which refers to New or Full Moon only. Most syzygies will occur with the Sun closer to the node than 7°.)

When Full Moon coincides with the Moon being opposite the Sun in the sky, the Moon must pass into the shadow of the Earth which, of course, must also lie in the plane of the ecliptic. The Moon is then eclipsed. If the Sun is at its maximum distance from the node during the half lunar month, the Sun is only partially obscured as seen from the Moon, and the eclipse of the Moon is only partial—an event that can pass unnoticed by the casual observer. When the Sun and Moon approach the same node, the Moon's shadow in space is directed towards the Earth and, from places on the very narrow track the Moon's moving shadow makes across the surface of the also moving Earth, a total eclipse of the Sun is seen.

By a strange coincidence, the Moon and Sun appear to be about the same size in our skies. In fact, the Sun is just over 400 times the size of the Moon, but it is also, on average, 390 times further away from Earth, so that generally it is only a very small fraction larger than the Moon in apparent diameter. But, due to the ellipticity of the Earth's orbit, the Sun's apparent diameter changes from a maximum in January to a minimum in July, while the Moon's apparent size also changes, due to the elliptical nature of orbit around the Earth. In the Moon's case, its closest approaches to Earth (perigees) do not recur at the same part of the sidereal month since perigee moves over 3° eastwards, and so happens about 5½ hours later, each month.

Depending upon the relative sizes of the Moon and Sun, a total eclipse of the Sun can last for as long as 7 minutes 58 seconds if the middle of the eclipse occurs in July, when the Sun is at apogee (furthest from Earth) and the Moon is at perigee, and the eclipse is observed from the Earth's equator. Duration is usually much shorter than this, and, when the Sun's apparent size exceeds the Moon's, the eclipse is annular, the Sun appearing as a bright ring around the dark disk of the Moon at the mid-point of the eclipse. However, whether total or partial, eclipses of the Sun and Moon must take place each year. In some years there are as many as seven—comprising either four solar and three lunar or five solar and two lunar eclipses—and in other years there are as few as two total solar eclipses with only partial lunar eclipses. Since the Moon's shadow is so small by the time it falls on the Earth, the track of the total solar eclipse across the Earth's surface is very narrow. Any one place has a chance of being in the track of a total eclipse of the Sun only once every century or two.

The ancient Egyptian astronomers noticed, over the hundreds of years that they recorded astronomical events, that eclipses of the Sun and Moon followed a sequence which takes approximately 18 years and 11 days to complete. This is because one node takes just

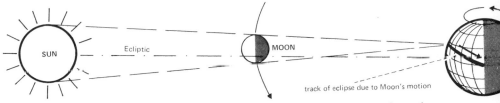

The principles behind an eclipse of the Sun. The diagram is *not* to scale; as drawn it would produce a partial eclipse of the Sun over the entire Earth. In reality, the visibility of even the partial eclipse is limited to certain parts of the Earth's surface. The spot on the Earth's surface produced by the Moon's shadow moves across the surface, due to the Earth's rotation and the Moon's motion combined. Note that the Moon has to be near to, although not exactly on, the ecliptic.

over 18½ years to move once round the ecliptic so that, almost exactly 18 years after one node has passed through a particular position, the opposite node will be at that position. At the same time, the Moon has passed through an exact number of lunations (sequence of phases, or synodic months) so that, if an eclipse takes place at a particular node, there will be another eclipse 223 lunations—or just over 18 years—later. But the time of day, and hence the place on Earth from which this total eclipse would be visible, will be different.

Each eclipse forms part of a sequence of eclipses in which the relative positions of the Sun, Moon and Earth gradually change due to the slow but regular change of such variables as the position of the Moon's perigee and all the other irregularities that have been mentioned. From any one place on Earth even partial eclipses of the Sun are relatively rare. Although eclipses of the Moon can be seen from any part of Earth where the Moon is above the horizon at the time of the eclipse, many years have no total lunar eclipses. In view of this, it is really quite amazing that the Egyptian astronomers were able to predict eclipses from their knowledge of this 18-year-11-day cycle, the Saros.

Secrets of the Saros were kept by the priesthood along with all the other practical astronomical knowledge, and gave the priests apparently amazing powers of prophecy and even apparent control of the heavens. The Sun God, they would predict, would be swallowed up because of the offences committed by the people. When the eclipse happened, the priests could promise to restore the Sun, on their terms! Such behaviour may be a more likely occurrence in fiction, where eclipses have been used by those in the know to confound primitive peoples who are threatening them at the time, but certainly the principles of an élite priesthood having knowledge of astronomical mysteries helped to reinforce the incredible belief in astrology which persists even to this day.

Stellar Parallax

There is another link between astronomy and time, also arising from the motion of the Earth in space, on the cosmic scale. This is the distance from Earth to the stars.

Each step in astronomy has to be founded on previous discoveries. This is particularly true in the scaling of the Universe which has, throughout the history of astronomy, tended to be a revision of previous ideas upwards: the Universe has always proved bigger than we previously imagined. This is likely to be true now as at any time. When Kepler formulated his famous laws, he provided the key to scaling the Solar System: once one distance was known, all the distances were known. Simultaneous observations from widely separated places on Earth of a body such as an asteroid, in orbit around the Sun, and comparatively close to the Earth, allow the distance to the asteroid to be calculated using the differences in its observed position and the known separation of the observers on Earth. The change in the asteroid's apparent position is due to parallax, the same effect that gives a change in perspective from one eye to another (look at this page through the right eye only, then the left only; the words will appear to move) or causes relative differences in the speed with which objects in the countryside appear to move past as seen from a train.

One of the best known applications of parallax measurement is taking bearings at sea of objects on the land to determine the ship's position. As the ship moves at a known speed over a known interval, and hence moves a known distance, the relative bearings of some fixed object on land can be compared, giving a fix of the ship's distance from that object.

The same principles can be used, taking the diameter of the Earth as a baseline, for comparatively "local" measurements, or using the diameter of the Earth's orbit, some 300 million kilometres, for more distant observations. In all cases, the angles concerned are very small, and hence accuracy of timing is vital. This was the sort of problem that led astronomers to adopt Ephemeris Time instead of Universal Time.

If one observes the position of a nearby star (once it is known or suspected that it is relatively close) against the background of the very distant stars on two separate occasions about six months apart, one is seeing it from two positions in space some 300

The telescope is perhaps the single most important chronometric instrument at Man's disposal; here are two of the most notable: (*above right*) the McMath Solar Telescope at Kitt Peak; (*below right*) Lick Observatory.

million kilometres apart. (In fact, the Sun's motion in space around the Galaxy must also be taken into account. Strictly, the Earth's path through space relative to the stars is not an ellipse, but an elliptical spiral around the path of the Sun.) The distance the Earth has changed in position between the two observations can be determined accurately, and together with the measured parallax this gives a distance to the star.

Unfortunately, this idea was fine in principle but, once again, the Universe proved bigger than expected. When the first parallax measurements were made by Bessel in 1838, the star he had chosen, 61 Cygni, had a parallax of only 0·31″ (less than one ten thousandth of a degree). This corresponded to a distance of 90 million million kilometres.

This distance was astonishing. Although the distance to the stars was thought to be very great, this star is so far away that the light sent out from its surface takes about ten years, travelling at 300,000km/s, to reach the Earth. We see 61 Cygni as it was ten years ago. If it suddenly exploded now, we would not know about it for ten years. Which brings up the question of what do we mean by 'now'?

61 Cygni is on our doorstep, as stars go. Most stars are far too distant for us to measure their parallax by this direct method, but the dynamic parallax due to the Earth's motion provided an independent method of checking other means of deducing the distance of stars. As these techniques developed, it became clear that we see the stars as they were years or more often hundreds or thousands of years ago. Using telescopes and radiotelescopes, we can see the Universe as it was millions or even billions of years ago.

Astronomy provides the most obvious examples of the relativity theory that Einstein put into mathematical terms in the early part of the twentieth century, and which will be examined in Chapter 5. What relativity means, in the most simple of terms, is that there is a different "now" for every observer, but that this difference can be ignored at "normal" velocities and "normal" distances.

Coda

Through astrology and, later, astronomy, Man has learned a concept of time that pervades his daily life and affects the most profound of his philosophies and sciences. The Earth's daily and seasonal variations produced evolutionary effects in the animal and plant kingdoms, manifest in humans, for example, in an apparently in-built "biological clock"—see Chapter 4. Such an effect on evolution enables us to predict that, if we discovered another inhabited planet circling some distant star, we could deduce a great deal about the living creatures on that planet merely by observing its period of rotation, the time it takes to circle its primary star, and the inclination of its axis to its orbit.

It is as well to remember that we are children of planet Earth, and our concept of time is inextricably bound up with our daily and yearly experience of our planet. The physicist's question, "What is the nature of time?" may seem meaningless to many of us, but we have already seen one way in which we might question the meaning of "now", and later we shall see others.

It is interesting to speculate on what our concept of time would be like if we were one of those thinking clouds of matter adrift in intergalactic space that have appeared in a number of science-fiction stories. We, the cloud, would have no idea of life or death since we would be adrift from our kind (if there *were* any more like us). We would be aware of certain physical phenomena, such as the redshifts in the spectra of those galaxies that appear faintest and hence are probably the most distant from us, but linking that phenomenon to a velocity would be more complicated for us than for planet dwellers whose lives are dominated by the problems of getting from one point to another at the appropriate speed, and hence at the appropriate time. But let us, the cloud, harness a little energy to set ourselves revolving slowly so that the galaxies wheel slowly past our "eyes" and we can count the transits of our brightest galaxy. This galaxy is near enough for us to examine its structure and appreciate its beauty, so we rather lose interest in the other faint points of radiation. Over a period (seconds, or millennia?) we become used to this periodic nature of our existence and we may start thinking that there is something special about time. . .

Plato said that time and the heavens came into being at the same instant in order that, if they were ever to dissolve, they might be dissolved together. Such was the mind and the thought of God in the creation of time. Perhaps time is something only beings on a spinning planet like the Earth, with a clear atmosphere, think about at all. It is impossible to relate time with conciousness alone. As John Donne put it in *The Sun Rising*:

> *Love, all alike, no season knows, nor clime,*
> *Nor hours, days, months, which are the rags of time.*

RAK

3 From Sundial to Atomic Clock

One of the yardsticks by which Man's rise from barbarism to present-day civilisation may be charted is his increasing desire to measure the passing of time to ever finer tolerances, to catalogue the years and to know the precise time of day whenever he wishes. For early Man—the prehistoric cave-dweller—hours, minutes and seconds were of neither interest nor comprehension. He would get up with the dawn, spend the day doing just enough hunting and scavenging to enable his family group to subsist, and would bed down for the night at dusk. Although doubtless aware that the Sun's high point split the day in two, he did not attempt to measure the progress of the day until long after he had begun to chart the aggregations of days; in other words, calendars came before clocks.

From a very early period he noted the passing of time from various natural phenomena: from day and night, the lunar cycle, women's menstrual cycles, the seasons of the year (harvest, snow-fall, leaf-sprouting, plus the less regular wet season, drought, river flood, animal migration, and so on), and stellar patterns (the year can be derived from the cyclical progression of constellation positions across the sky). Thus the day, month and year are all natural, easily observable time-divisions. It is obvious that they were noted and even marked (perhaps by religious ritual) much earlier than any surviving records indicate. Time was also measured from, or by reference to, less frequent occurrences, such as generations, or some remarkable happening (forest fire, flood, eclipse or plague, for example).

Early Calendars

To state that the year is "easily observable" is, of course, a generalization—a sweeping oversimplification. It is easy enough to be certain of the length of a year to within a few days, from the repetition of a particular stellar position at dawn. Indeed, a South

For the ancient Egyptians, Sirius' first morning rising occurred on a different date every four years because they never added a day to their calendar year of 365 days. This extract from a calendar of Pharaoh Thutmose III (1501–1447BC) assigns the feast celebrating the star's first appearance to the "28th day of the third month of summer".

American Indian tribe has been discovered within the last couple of decades which still calculates its year-end from the setting of the Pleiades (that tight group of seven naked-eye stars in the shoulder of the constellation Taurus) immediately before dawn, and they use the same word for both "year" and "Pleiades". On the other hand, if one attempts to keep account of the passing of years by counting the days, one is confronted by awkward fractions whether working from stellar or from lunar data. The solar year is slightly less than 365¼ days, while the lunar year (twelve lunar months, each of 29½ days) is only 354 days. The devising of a calendar

An elementary method of recording the passage of time in a lunar month was the string calendar. An account of the days was maintained by threading string through each of the thirty holes.

sufficiently exact to take account of these odd numbers without getting out of step with the seasons was not achieved until the sixteenth century and not universally accepted until the twentieth.

All early civilizations possessed some form of calendar, though the lengths of month and year differed greatly. The shortest was the six-month year adopted by some tropical peoples. This contained a wet season and a dry season—an obvious cycle. At one stage the ancient Babylonians also used a six-month year, based on lunar eclipses. The tally of days and years was frequently the responsibility of the priesthood, because it was primarily for religious purposes that any count was kept at all.

The earliest calendar system was almost certainly the work of the Egyptians. They based their year on stellar observation—specifically, on the rising of Sirius, the Dog Star, with the Sun, which marked their new year. They determined the interval between two such heliacal risings as 365 days, which were divided into twelve thirty-day months with the extra five days added on at the end of the year. After only a few years it became obvious that this period was slightly short, with Sirius heliacal rising and the procession of the seasons all coming later and later in the official year. But, because tradition is always easier to keep than to break,

The annual flooding of the Nile and the beginning of the fertile season was heralded in ancient Egypt by the first appearance of Sirius in the morning sky. This second-century BC relief depicts the cow-headed goddess Isis watering the sprouting corn.

and because few people outside the priesthood would have known the official date in any case, the system was maintained over the centuries, with the calendar and solar years growing more out of phase with each other. Eventually, after 1,460 solar years (365 × 4) new year's day again coincided with the heliacal rising of Sirius. Then the system was allowed to creep out of step again . . . and again. We know that there was a convergence in the year AD139. Therefore another convergence must have occurred 1,460 years earlier, in 1321BC, and before that in 2781BC and 4241BC. The starting point of the Egyptian calendar system was in one of these last-mentioned two years, probably in 4241BC.

In ancient Babylon and Assyria there were forms of calendar in use before 1900BC. After an early period in Babylon when a six-month year based on lunar eclipses was tried, a year of twelve lunar months (354 days) was introduced. A thirteenth month was intercalated (inserted into the calendar between two existing months) from time to time, when it was felt to be necessary. There were different month-names in use in different areas until the unification of the states (by 1600BC). Eventually, by the fifth century BC, the Metonic cycle was adopted, which brings the solar and lunar years into step by intercalating seven months in nineteen years.

The Assyrians based their system on heliacal risings and settings, initially (by 1900BC) having a year of 360 days divided into twelve equal months, to which was added a period of fifteen intercalary days (half a month) every three years. Later, other systems were introduced, including the Babylonian one after Assyria had conquered Babylon.

Other Mediterranean peoples with calendar systems of great antiquity are the Jews and Greeks. The Jewish system has changed tremendously over the last three to four thousand years, with years of varying lengths and bases. Calculation has always been a complex process—too much so to go into details here.

The first ancient Greek calendars may date from the Trojan War period (c. 1200BC). Many of the city-states had their own indigenous systems, although their month-names were often similar (but did not always refer to the same month). Year-ends varied from the first new moon after the summer solstice (Athens) to the autumnal equinox (Macedon and Sparta), although the Egyptian out-of-step calendar was favoured by some. It has been said that the Greeks did not worry overmuch about the accuracy of their calendars, yet it is known that many Greeks possessed their own peg calendars (*parapegmata*) with which it was possible to find the positions of the lunar months, the equinoxes and so on. It was the Athenian astronomer Meton who, in the fifth century BC, invented the Metonic system of intercalation which we have mentioned, and which is still a feature of the present-day Jewish calendar.

It was not only in the "cradle of civilisation" on the borders of the Mediterranean Sea that early calendars were developed. In Asia and America, too, there existed quite sophisticated systems in prehistoric times. The earliest proven of these were the essentially similar calendars constructed by the Maya and Aztecs in Central America by about 3000BC. These were based on a 260-day cycle—a progression of twenty names operating alongside the numbers

POP UO ZIP ZOTZ TZEC XUL

YAXKIN MOL CHEN YAX ZAC CEH MAC

KANKIN MUAN PAX KAYAB CUMHU UAYEB

The months of the Mayan calendar, showing eighteen months each of twenty days plus a nineteenth, Uayeb, a five-day period.

1-13, although the names varied between areas and periods. Although this cycle was important for religious purposes and for prophecy, it was linked with a 365-day solar year composed of eighteen named months, each of twenty days, plus Uayeb, a year-end period of five "evil days". This gave a 52-year progression with different day-month combinations for each of its 18,980 days, the period being known as the Calendar Round. There is some evidence to show that the Maya were able to calculate the accumulated error between their 365-day year and the actual— over a period of some four thousand years—although they did not correct their calendar. Certainly the Maya had a highly developed time sense, being able to think in terms of an almost limitless past and future. They possessed words for high numbers (increasing by magnitudes of twenty) up to 3,200,000 (20^5), the *kinchiltun*, and they began a Long Count of time in the year 3113BC. The Aztecs operated a similar system, although with different names. There is no evidence that they began the Long Count as early as the Maya, but certainly they calculated time from the same base.

The Hindu calendar dated from about 1500BC and was based principally upon lunar months, although with frequent corrections and intercalations to ensure that it did not run ahead of solar time. There were twelve named months, divided into six pairs for different seasons (spring, hot season, rains, autumn, winter, dry season) and the calendar's main purpose was religious. (It was also partly cosmographical.) The other leading religious groups of the

Indian sub-continent—the Brahmans, Buddhists and Jains—used rather similar calendars.

As might be expected, the Chinese had a complex system of calendar calculation in operation from a very early date. This was probably as early as the eleventh century BC, but later manuscripts claim that it originated in the twenty-seventh century BC. Its most important feature was a sixty-day period of named days. From at least the seventh century BC a twelve-year cycle was used, based on Jupiter's passage through the constellations. Since this takes in fact only 11.86 years, occasional alterations were made to keep the years in step with Jupiter's observed position. Running parallel with all this were the fixed points of the equinoxes and solstices, so frequent intercalation was employed.

The Romans, too, had an early system—the republican calendar—which was in use from the time of Numa Pompilius (the beginning of the seventh century BC) until its reform into the Julian calendar under Julius Caesar in 46BC. The republican calendar was based on lunar months and had 355 days. It was complicated by varying month lengths, by occasional intercalary months and, at times, by political intervention, so that by 46BC it had become three months out of step with nature. The Julian calendar was an important advance over everything which had gone before because it was the first to be based on a year of 365¼ days, with provision for an intercalary day to be placed in February every fourth year. (February was originally the last month of the year in the republican calendar, although this had been altered to the present-day order more than a century before Caesar's reforms.) The Julian calendar survived as the major European system for over sixteen hundred years before being replaced by the Gregorian calendar.

While dealing with early calendars one must mention the calendrical nature of certain megaliths. The circles of standing stones which form Stonehenge, Wiltshire, are solsticially aligned. In other words, they point approximately towards the point of sunrise at the summer solstice (about June 21). Various significant lunar and stellar alignments have also been ascribed to the stones, either separately or in groups, although stellar positions have altered considerably over the 3,500 years or so since Stonehenge was built, while a number of the stones have been moved or replaced, so these alignments are less certain. Similar solsticial alignments are to be found among the megaliths in Brittany, particularly at Carnac, where stellar alignments have also been proven, and also at the fine circle of standing stones near Callanish on the island of Lewis in

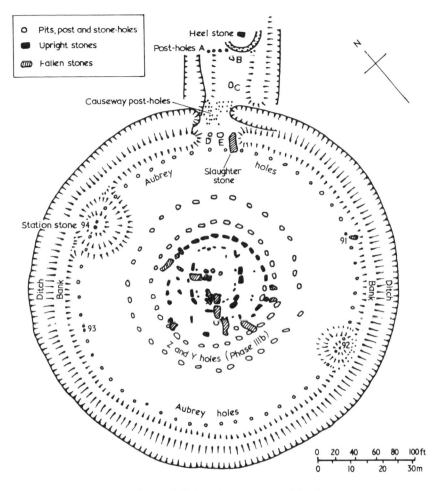

Heel stone
Post-holes A
B
C
Causeway post-holes
D E
Aubrey
Slaughter stone
holes
Station stone 94
91
Ditch
Bank
Ditch
Bank
93
92
Z and Y holes (Phase IIIb)
Aubrey holes

0 20 40 60 80 100ft
0   10   20   30m

A plan of Stonehenge, the world's best known megalithic observatory.

the Outer Hebrides. On the other hand, there is precious little evidence of any such significance in the case of the Great Pyramid (the Pyramid of Khufu or Cheops) at Giza, near Cairo, yet it is claimed by some writers to have been built partly as a guide to the equinoxes and a forecaster of eclipses. This is strongly denied by the experts.

Two connections which the ancients made between time and space require a mention here: the Long Year and the Harmony of the Spheres. Several early civilizations arrived independently at the concept of the Long (or Great or Perfect) Year, being the length of time between the heavenly bodies (Sun, Moon, planets and stars) being in a particular position and their return to that position. The astronomers of these civilizations all realized that many thousands of years would need to elapse, and this shows an early ability to think in terms of very long time-periods. The ancient Egyptians calculated the Long Year as 30,000 years. In the eighth century BC the Babylonians observed that the stars altered their annual positions at the rate of one degree every seventy-two years and would thus take 25,920 years to return to their original places. One must remember that geocentric theories of the Universe were generally accepted at this time, so that the idea of the parade of fixed and wandering stars returning to an exact formation at lengthy but fixed intervals—and, in particular, returning to the formation existing at the Creation—was easily believed. It meant that the Universe was a regular, dependable construction, demonstrating the Hand of God (or gods). Plato estimated the Long Year at about 26,000 years, and the Hindu astronomers chose a similar figure. The Arabs, at a later date, plumped for 49,000 years.

According to the ancient Greek geocentric theory of the Universe, the Earth was at the centre of eight concentric spheres. Seven of these held the "planets" (including the Sun and the Moon, but excluding Uranus, Neptune and Pluto, which were not known then) and one the fixed stars. It was Pythagoras, in the sixth century BC, who, having discovered that all musical intervals are subject to certain mathematical ratios and depend upon the frequency of vibrations, deduced that there must be numerical order and a musical scale in all things—in Heaven and Earth. Since the planets moved at different rates their motions would obviously produce sounds according to these rates, and such sounds would be bound to harmonize. So the idea of the Harmony of the Spheres originated. Although Kepler disproved the geocentric theory at the beginning of the seventeenth century, he perceived a new harmony of the planets in their relative distances and velocities.

Weeks and Hours

Although the day, month and year are natural time divisions, the week and the hour are arbitrary periods, not met with in nature but fixed by Man. Different men, at different times, have set varying values to their lengths.

To the most primitive peoples there was no need for such a thing as a week, but with increasing growth of culture two reasons arose. One was the need for a regular day to be set aside for religious devotion. The other was commercial—there had to be a fixed and fairly frequent market day. Weeks of all lengths have been used somewhere at some time, from four days (in West Africa) to ten days (by the Incas of Peru). The seven-day week (possibly favoured because of the supposed magical or religious importance of the number seven) has been observed from the earliest times by the Jews and, from the advent of the Julian calendar, by the Romans. Throughout recent history arguments have been advanced for altering the seven-day week to a longer or shorter cycle. (H. G. Wells advocated a ten- or eleven-day week with a three- or four-day weekend.) None have been taken seriously enough to be put into general practice.

Nor did primitive Man need to divide his day into hours until he was both advanced enough to construct some means of measuring the passing of those hours and civilized enough to accept the convention of keeping appointments—for commercial, religious or, indeed, epicurean purposes. The hours into which the day was divided were extremely varied in number and length. If one is keeping track of time by means of the Sun and dividing the period of daylight into twelve equal hours (as the ancient Egyptians did) the length of the hour will vary with the seasons and will be equal to the standard hour which we know on only two days in the year—at the equinoxes. To live by a system of hours of changing length may seem complex and unnecessary to us, but it is a system which prevailed for several thousands of years among widely scattered civilizations. "Temporary hours" is the name given to these hours of varying lengths. The Babylonians had a day and night of six temporary hours each, while the early Hebrews made do with half that number. The Japanese used temporary hours right up until the nineteenth century: their mechanical clocks were adjusted by a "clockman" every two weeks, or else built to show the different hour-lengths. Also, for religious reasons, none of their hours could be called 1, 2 or 3, so their six hours each of day and night utilized the numbers 4 to 9 twice, once in a rising sequence and

The time stick of a Tibetan priest.

once in a falling sequence. By contrast, the Chinese had achieved an equal-hour system, with a day and a night of twelve hours each, by the fourth century BC. In Europe temporary hours were in general use until the fourteenth century, which meant that the hours of daylight (which can vary from eight to sixteen per day in British latitudes) were always divided into twelve until then.

Non-Mechanical Timekeepers

Primitive Man must have noticed that the Sun cast a shadow of varying length. He would have used the shortening of his own shadow (or that of a tree, perhaps) as an indication of the approach of noon, and the lengthening of shadows as a warning of sunset, by which time he would need to be home. The varying direction of the shadows and their lengthening and shortening not just day-by-day but in accordance with the seasons must also have been noticed. At some stage, presumably several thousand years BC, someone first thought to calibrate this varying shadow so as to divide up morning and afternoon. It may have been a priest, anxious to know when to prepare for a particular ritual, or perhaps a farmer or craftsman who needed to do a certain job before sunset and wished to check on his progress. The method of calibration was to set up a thin vertical stick or pillar (a gnomon) and to inscribe a series of arcs or lines on the ground to indicate shadow-length at dawn, noon, dusk and a few intermediate points. The positions of the

Egyptian shadow-clock dating from between the tenth and eighth centuries BC.

lines would need to be altered every month or so, to cope with the changing angle of the Sun.

A shadow-clock of this simple type was Man's first timekeeping device. The principle was certainly known to the Chinese as early as 2500BC. A fairly sophisticated type of instrument from Egypt, *c*.1000BC, is the earliest to survive. The only disadvantage of this type, which was about twelve inches long and made of stone, was that it needed to be aligned with the calibrated bar pointing towards the west in the morning and towards the east in the afternoon. No example with a T-piece at each end is known to exist. There are two biblical references to sundials, in the second book of *Kings* and in *Isaiah*. These mention the same incident, which experts have placed in the year 741BC. In *II Kings*, 20:11 it says, "And he brought the shadow ten degrees backward by which it had gone down in the dial of Ahaz."

In about 340BC the Babylonian astronomer-priest Berosus developed a more complex and accurate model—the hemicycle. This was a block of wood or stone with a hemispherical depression and a gnomon in the centre, its top flush with the top surface of the block. Around the curved surface was engraved a series of arcs each divided into twelve equal parts to signify the hours of the day.

The Greeks developed the sundial further, experimenting considerably with hemicycles and with flat dials set at various angles. All the citizens (as opposed to the slaves) of the city-states had sundials and were able to keep a fairly consistent common time—in temporary hours, of course. There were many public sundials consisting of tall columns. Hence the line in Aristophanes' play *The Frogs* (*c*.405BC) which runs: "When the shadow is ten steps long, come to dinner." The Tower of the Winds in Athens (built in about 100BC) is octagonal with a sundial on each side, and it was partly filled with water, for it was a water-clock, too. Strictly speaking, all these instruments were shadow-clocks rather than sundials, because they told the time by the length of a shadow, not its direction.

The Romans captured a sundial from their near neighbours, the Sammites, in 290BC, and thereafter made great use of such devices.

In AD600 Pope Gregory I ordered sundials to be placed on all churches. This practice continued for a few centuries, and many pre-Renaissance churches in the UK and continental Europe still have a sundial on their wall. This is, most often, of the simplest type—the scratch dial—with a horizontal gnomon projecting from the wall and hour-lines scratched into the wall below it.

It was the Arabs who further developed sundials. Abdul Hasan, at the beginning of the thirteenth century, wrote a treatise on the construction of hour-lines on surfaces of various shapes. He is credited with introducing equal hours to the West, although only for astronomical purposes.

When the earliest mechanical clocks came into general use, in the late fourteenth century, it might have been expected that sundials would quickly fade away and be forgotten. But this did not happen. Clocks were expensive, quite large, not very accurate until the seventeenth century, and inclined to signify their dislike of travel by stopping whenever moved. Consequently, once sundials had been converted to equal hours, they held several advantages. Their design and manufacture flourished during the Renaissance, when the variety and ornamentation increased dramatically. The angled gnomon appeared, often ornamented with wrought-iron scroll work, and set to the same angle as the latitude in which it was to be used. The dials were not only engraved with the hours but also with intricate designs and cheerless mottoes such as "Hours are time's shafts, and one comes winged with death" and "As the long hours do pass away/So doth the life of man decay". This type includes most of the well-known garden sundials which survive today; the dial is normally horizontal, but occasionally vertical.

Another type worked on quite a different principle. A hole was cut in the roof of a building (most often a church) through which a shaft of sunlight fell to illuminate a portion of the dial marked on the floor below. At least one such is still in operation, in Milan Cathedral.

Particularly popular during the Renaissance were sunken dials. Generally made of stone and based on the hemicycle, these were often incorporated into the walls of houses, or placed on church buttresses or tombstones.

The tablet dial made its appearance during the fifteenth century. It was portable, composed of two rectangular pieces (or tablets) of wood or metal hinged together at one side. These were opened to more than 90° to reveal the hour lines on their inner faces, and the gnomon was a piece of taut string. Carried by travellers, the tablet sundial became very popular in the sixteenth century.

More compact was the ring dial, only about three inches in diameter, with a pin-hole in its wide rim through which the Sun's light shone to illuminate the hour-line. These dials, introduced in the late seventeenth century, were adjustable for any latitude or

A garden sundial, of the type that we have all seen.

A typical example of a tablet sundial.

A portable universal ring sundial of the type patented by H. Sutton in 1660.

A tremendous step forward: Oliver's mean-time sundial, patented in 1892.

season. They were much more sophisticated than earlier ring shadow-clocks, actually worn on the finger, with a folding gnomon.

In the late eighteenth century Catherine the Great of Russia had sundials set up on milestones along the road between St Petersburg and Moscow. In the nineteenth and twentieth centuries the descendants of the sundial—the solar chronometer and universal heliochronometer—were developed for accurate scientific measurement.

Sundials or shadow-clocks are useful during the hours of daylight in sunny climates (their greatest use was in the Mediterranean lands), but what about night-time and cloudy days? From very early times there were alternative systems of timekeeping which did not rely on the Sun.

Most widespread was the clepsydra or water-clock, which was certainly used in Egypt before 1400BC. Frequently it consisted of a metal or earthenware pot with a small hole in the bottom, through which water escaped at a predetermined rate. The level of water remaining gave the hour, which could be read off a scale cut into the side of the pot. An alternative approach involved floating an empty holed pot in a larger, water-filled one and telling the time by

A cast of an Egyptian waterclock, or clepsydra, which dates from the second century BC.

A Chinese incense clock, a device which measured the passage of time using the rate at which incense burned.

An unusual sand-clock calibrated in minutes.

the rate at which the smaller pot filled with water. Or an unholed pot could be filled by drips from an overhead reservoir.

None of these systems was particularly accurate and, in any case, the scale of hours had to vary with the seasons if it was not to clash with the changing temporary hours of the shadow-clock. Nevertheless, the clepsydra spread right across Europe, being used by the Greeks (to limit the length of speeches in the Athenian courts, among other purposes), Romans, Saxons and Arabs. It was separately developed by the Chinese. One thirteenth century clepsydra was filled with mercury to avoid freezing during the winter, on the orders of Alphonso X, King of Castile.

Other timekeeping systems involved the flow of sand through a small aperture (the hourglass) and the burning of graduated candles or ropes, or of measured quantities of oil or incense. The sand-clock or hourglass achieved considerable accuracy and was in use for many centuries. (One can afford to ignore its minor three-minute rôle today.) It may first have been used by the Roman army to measure the length of night watches, though none remains as proof.

The art of glass-blowing was rediscovered by an eighth-century monk at Chartres who made a sandglass, and those in existence date from that century. Sandglasses were made in various durations, including ¼ hour, ½ hour, 1 hour and 4 hours. Apart from the very earliest, all were double-ended and could be inverted. The four-hour models were in regular use aboard ship to measure watches right up to the late eighteenth century, when accurate ship's chronometers were first built.

One other device of some interest was the nocturnal or night clock, designed for telling the time from the positions of the Pole Star and the Great Bear constellation. It could be set for the season of the year and was made in the sixteenth century.

Early Mechanical Clocks

It is not possible to say when the first true mechanical clock was constructed or by whom. Confusion has been caused by the impre-cision of early manuscripts which refer to "horologes" prior to the fourteenth century, but this was a blanket term, covering sundials and waterclocks as well as those driven by weights.

The word clock is derived from the Latin *clocca*, a bell, and it is believed that the very earliest mechanical clocks were designed to strike a bell at one-hour intervals rather than to show the time con-tinuously by means of a dial and hand. But in monasteries for a

A ship's four-hour sand-clock, probably from the eighteenth century.

A brass nocturnal from Cologne, signed "Casper Vopeli' Faciele".

couple of centuries prior to the development of mechanical clocks it was the duty of one monk to watch an hourglass or water-clock, or even to read aloud a pre-timed passage from the scriptures, and ring a bell each hour; this practice resulted in the mistaken belief that monks in the twelfth or thirteenth centuries (or even earlier) had invented the weight-driven clock.

Further confusion originates from the automata—generally water-powered—which were built by the Arabs as early as the ninth century. These were occasionally extremely elaborate, such as the celebrated water-clock presented to the Emperor Charlemagne in AD809 by the Caliph of Baghdad. It struck the hours (by means of falling ball bearings) and re-primed its own striking mechanism, but it was not a mechanical clock.

The earliest authenticated weight-driven clock was one in Milan, in the palace chapel of the Visconti. That struck the hours and was built in 1335. Fairly definite references antedate this, such as the mention of a clock by Dante Alighieri in his *Paradiso*, which was completed in about 1320, but no reliable descriptions remain. Yet it is obvious that weight-driven clocks with escapements were being developed throughout the first half of the fourteenth century, because quite a number of public clocks were in existence in various parts of Europe by the year 1350, and those that remain show a certain complexity which must have taken decades to develop.

The very earliest mechanical clocks were driven by a principle which was in common use for five hundred years and is still occasionally used in clocks today—the energy of falling weights. These are suspended by cords wound around a drum, which is attached by a wheel to the clock's train of gears. To prevent the rotation of the drum and gear train from accelerating and using up all the

A Japanese silk painting from the eighteenth century showing a woman winding a clock. This clock depends upon the driving force of falling weights, a principle which was used in the western world for about half a millennium.

energy of the weights very quickly, a mechanism known as an esca-pement is employed. The earliest European escapement—its inventor is unknown—is the verge or crown wheel escapement. An arbor leading from the gearing (and forced round by the falling weights) makes the crown wheel rotate, but the pendulum arbor has two pallets fixed to it which alternately block and release the wheel as the pendulum swings, allowing the wheel to turn slowly in a series of forward jerks. The wheel also provides the motive power to keep the pendulum swinging. (But note that the pendulum was not used to control the clock's pace by providing a steady one-second beat; such a pendulum was not developed for another three hundred years and more.) The crown wheel could be mounted either horizontally or vertically. If the latter, it was operated by a foliot balance. The balance rotated back and forth, suspended by a thin cord, and the weights on the balance arms could be adjusted to make the clock go faster or slower (although there was no regular period of swing or rotation and the regulation of these early clocks was difficult).

The existence of an ancient Chinese escapement must be men-tioned in passing. This is described in a Chinese manuscript of about 1088, as part of a clepsydra. A cup was slowly filled by a water-drip. When its weight reached a certain value it would tip a balance, permitting a chain of gears to advance by one notch, and also emptying the cup so that it could be refilled. Each cycle would appear to have taken about a quarter of an hour, although no working example has ever been found. It is almost certain that this was unknown in the West, so that the verge escapement was a parallel development.

In England the oldest surviving clock is that of Salisbury Cathedral, dating from about 1382. But this has been modernized at intervals over the centuries and now contains little of its original substance.

The clock of Wells Cathedral, made about ten years later, is also still in existence. It is now in the Science Museum, London.

All these early clocks were handmade from wrought iron by blacksmiths or locksmiths—an arduous task. A notable exception was the Dondi clock, of which the maker's detailed description, dated 1364, survives. This was made of brass, bronze and copper, with seven dials showing lunar and planetary motions as well as the time on a 24-hour dial. Its maker, Dondi, was Italian and the clock took, reportedly, sixteen years to construct, being obviously far ahead of its time.

A Chinese astronomical clock tower dating from 1088.

(*Left*) The striking mechanism of the clock from Wells Cathedral. (*Right*) An original illustration to the maker's description of the Dondi clock, dated 1364.

Very few clocks were cased until the sixteenth century—though cases were almost invariably of iron, too—and most of those constructed in the fourteenth century were turret clocks; i.e. were meant to be placed in a turret and to strike the hours as a public service, rather than being for domestic use. The earliest examples tended not to have a face or hands, although a few domestic clocks with hands (one per clock; minute hands were not generally fitted until the seventeenth century) were made in the fourteenth century.

The vagaries of the verge escapement and foliot balance system, plus the difficulty of working metal without sophisticated measuring tools, meant that accuracy was not very great; errors of fifteen minutes per day were common. But mechanical clocks were, even so, much in demand as status symbols among the rich and powerful, although no more useful than the sundials and sandglasses by which they were checked for accuracy. The clockmakers of Europe did their best to make improvements, as well as trying to satisfy the demand for their products. Although the Italians (including Dondi) were the best of the very early makers, some of the towns of southern Germany, particularly Nuremberg, Ulm and Augsberg, soon became famous centres with their own guilds of clockmakers.

Seeking greater timekeeping accuracy and an alternative to weights as the motive power, it was probably a Nuremberg clockmaker, Peter Henlein, who first built a spring-driven clock in about 1510. (There is some dispute over this; evidence suggests that spring-driven clocks were being designed in Italy in the 1470s, although no examples from that early date have survived, while Leonardo da Vinci includes a sketch of a fusee mechanism—a clock-spring force equalizer—in his notebooks, dating from about 1485—90.) The great advantage of a spring power clock over one with weights is portability; and it enabled smaller, neater domestic clocks to be built—as well as watches, which first appeared shortly after 1510, although whether the first were made in Germany or in Italy is uncertain.

A problem to be overcome was how to obtain a uniform force (or torque) from the spring, which would naturally exert a greater pull when fully wound up than when almost run down. Two solutions were devised, the fusee and the stackfreed. The fusee was the more important and long-lasting invention; it is still used almost unchanged today in some finely made clocks. The spring is housed inside a small barrel-shaped arbor, and it pulls cord or fine chain from the fusee, a spindle of gradually increasing diameter. The idea is that when the spring is fully wound its pull is against the smallest diameter, and as it runs down it gains greater leverage by pulling against a progressively wider diameter of spindle, so equalizing the force transmitted to the gear train.

The man normally credited with the first practical application of the fusee is Jacob Zech (or Czech) of Prague. A clock of his, incorporating the fusee, dates from 1525. (A clock with a fusee is known dated 1504, but the authenticity of this date is questionable.)

The inventor and initial user of the stackfreed are unknown. This device utilizes a second spring, which acts as a brake of gradually lessening effect upon the action of the mainspring, so equalizing the force of the mainspring. This is achieved by the second spring pressing against a snail-shaped piece of gradually decreasing diameter, so that the principle is the same as in the fusee. The stackfreed was slightly less efficient than the fusee but remained in use, particularly among the German clockmakers, for many years.

The advent of the spring-driven movement meant that relatively small, portable clocks became common. These were often (during the sixteenth century) drum clocks, so named from their shape. Generally the dial was on the upper face, and they tended to be

A sixteenth-century drum clock.

very ornate—many with striking and alarm mechanisms, some with astrolabes incorporated, and all embellished with gilt or handsomely cased. It must be remembered that although these were portable and would probably keep going in any position, they were not in the same category as pocket watches; typically they measured something like nine inches in diameter and five inches in height. Miniature versions were made in the same shape (canister watches), perhaps two or three inches in diameter and an inch or so in height, but these required the greatest degree of workmanship and were far less common. The cover of the case had to be lifted in order to read the time. Most were intended as travelling alarms, and have either an alarm or a striking mechanism fitted—occasionally both. The cases were normally pierced around the edge with many tiny holes to allow the miniature bell to be better heard. These were the playthings of the rich.

By the late sixteenth century upright spring-driven clocks were being made. They were domestic table clocks, often beautifully ornamented.

An innovation of about the year 1600 was the inclined plane clock, driven directly by gravity. Such a clock would take, for example, twenty-four hours to roll very slowly down its inclined plane, which would be about three feet long. This was its only motive power. The hand was fixed and showed the time against the revolving dial set into the side of the case.

It is important to remember that during the first three hundred years of mechanical clocks—up to about 1650—they became gradually more intricate and more highly decorated but gained little in accuracy. There were relatively few clock- or watchmakers in England until the late sixteenth century, and few examples of their work remain. But the craftsmen of continental Europe produced many fine pieces during this period (up to 1650), building them in the shapes of drums, statues, towers, books and globes.

Calendar Reform

Before going on to the next great advance in clockmaking—the pendulum—this survey must, to be chronological, describe the reform of the calendar at the end of the sixteenth century. The Julian calendar, dating from 45BC and taking 365¼ days as its standard for the year's length, had very gradually fallen behind the solar year. Although the discrepancy was small—about eleven minutes and fourteen seconds per annum, or one whole day every 128 years—it became noticeable over the centuries. By the eighth century AD the vernal equinox was arriving three days early each year, a circumstance noted by the Venerable Bede. By the thirteenth century the discrepancy amounted to just over a week, causing Roger Bacon to write his treatise *De Reformatione Calendari*. A fifteenth-century bid at calendar reform was frustrated when the astronomer appointed by Pope Sixtus IV to calculate the necessary changes was assassinated.

It was Pope Gregory XIII, in the late sixteenth century, who finally succeeded in setting the calendar straight. He gave his name to it, and the Gregorian calendar is the one we use today. The major alteration to the Julian calendar was the introduction of the rule that no century year (e.g., 1700) should be a leap year unless it was exactly divisible by 400, which meant dropping three leap year days every four hundred years. This sytem produces an error of only one day every 3,323 years (and a later amendment, to drop the leap year day for the year AD4000—if we should ever reach it—has reduced this error to one day in 20,000 years).

A more complex change brought about by the Gregorian calendar was the introduction of epacts (i.e., lunar-based calculations) to determine the date of Easter.

With the changes superintended by Christopher Clavius, the German Jesuit and mathematician (who has had a lunar crater named after him), the reform was adopted in most Roman Catholic countries in the year 1582. Protestant countries were slow to change. Britain adopted the Gregorian calendar as late as 1752 (when the omission of eleven days of September, to bring the country into line, resulted in protests and the cry "give us back our eleven days") and Russia only in 1917, after the Revolution.

The only serious attempt to break away from the Gregorian calendar and introduce a radically different system came in France under the First Republic. The Revolutionary calendar was introduced in 1793 (although indexed from the proclamation of the Republic in the previous year). The years had twelve months (all poetically renamed to fit the seasons) each of thirty days, plus five or six intercalary days. It began at the autumnal equinox (September 22), which was also the day following the Republic's foundation. After twelve years good sense prevailed over separatist zeal, and France readopted the Gregorian calendar from January 1, 1806.

More Accurate Clocks

At about the same time as Rome changed over to the Gregorian calendar, Galileo Galilei, as yet only a student in his late teens, was devising the idea of the pendulum as a clock-regulator by watching a swinging lamp suspended from a long chain in Pisa Cathedral. He later became a renowned scientist (infamous, even, for his denial of the Roman Catholic tenet of a geocentric Universe) and attempted to build a pendulum clock.

But the first successful pendulum clock and the general application of the pendulum as a time-controller did not come about for

A working model of Galileo's application of the pendulum to timekeeping.

A pendulum pillar clock with oriental markings, although probably not of oriental origin.

Thomas Tompion, possibly the finest and certainly the best known of all British watch- and clockmakers.

An example of Tompion's watches.

another seventy years. It was Christiaan Huygens, the Dutch astronomer-physicist, working from Galileo's initial ideas, who designed a pendulum clock in 1656.

Although Huygens' pendulum kept a steady beat, it did not beat at exactly one second. It was probably the English physicist Robert Hooke who, in about 1660, discovered that a weighted pendulum approximately 39.14 inches in length will beat at exactly one second. This became known as the Royal Pendulum, and it enabled the second to come into common usage as a measurement of time, and for minute hands and, eventually, second hands, to be introduced on clock-faces. (The derivation of the word second is from *second minute*; i.e., the second division of the hour by sixty.)

Certainly it was Hooke who invented the balance spring in about 1660. This is often referred to as the balance wheel or hairspring, and it has been an important feature of watches ever since. It led to them becoming round and fairly flat—suitable for being carried in the waistcoat pocket. A contemporary and friend of Hooke's was Thomas Tompion, perhaps the most outstanding of all English watchmakers, and it was he who put Hooke's discoveries into practice.

Still in about the same year (1660), either Hooke or William Clement, a London watchmaker, invented the anchor or recoil escapement, which was a great improvement on the verge escapement and is still in use (with only minor refinements).

The combination of a one-second pendulum and the anchor escapement made extremely accurate clocks possible for the first time (more accurate, indeed, than solar time, which varies slightly, so that mean solar time had to be introduced—see Chapter 2). In addition, it proved possible to design clocks which would keep going for much longer than the thirty-hour upper limit which had previously been the case. It was found that a heavy pendulum coupled with a very short swing (angles of 3° or 4°) gave the best results, so the small shelf or table clocks which had been popular were quickly replaced (in England, at least) by tall, upright clocks, with the mechanism and pendulum all encased in wood. This was the beginning of the longcase or grandfather clock, which gave rise to a great demand for cabinet makers able to suitably embellish them. Often seven or eight feet tall, they retained their popularity in England for more than two centuries. This was Oliver Goldsmith's "varnished clock that clicked behind the door". Its introduction, in the third quarter of the seventeenth century, coincides with the start of a period of pre-eminence for British clockmakers.

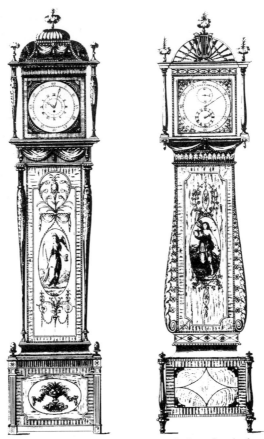

Line drawings showing Thomas Sheraton's two designs for clock cases, published in 1791. Great emphasis is placed on the use of veneers of exotic woods and marquetry work. No examples are known to exist made to these designs, although casemakers did on occasion use isolated "Sheraton" features.

Although these longcase clocks should have been exceedingly accurate, it was soon realized that on hot days in the summer they would lose. This was due to the expansion of the pendulum: an increase of length of one-thousandth of an inch makes the clock lose about one second per day. It was easy enough to fit a manually adjustable screw to the pendulum bob, but clockmakers searched for automatic methods of compensation. George Graham, who in 1715 had invented the dead-beat escapement, a refinement of the anchor escapement which avoided the problem of recoil, produced in 1721 a brass pendulum with a mercury jar set into it. If the brass expanded because of a rise in temperature, lowering the centre of gravity of the bob, so the expansion of the mercury up its jar would raise the centre of gravity by the same amount. This elegant solution is still in use today in some clocks, as is the dead-beat escapement. Another of Graham's achievements was the perfection of the cylinder escapement for watches in about 1725, enabling them to be made thinner and more accurate than before.

A different automatic compensator was invented by John Harrison, a Yorkshireman, in about 1725. This was the grid-iron pendulum, made up of nine parallel rods, alternately steel and brass. Harrison was responsible for several other clock innovations, including the delightfully named grasshopper escapement, in 1730, and a system for maintaining power in a clock during winding, in 1734. (A weight-driven clock normally stops during winding, but in certain cases, particularly with scientific or extra-

John Harrison (1693–1776), the British instrument maker whose series of marine chronometers provided at last an instrument capable of accurately measuring time at sea, shown here holding his Number 4 chronometer, which won him a prize of £20,000 from the British Government.

accurate clocks, it is important for them to keep going smoothly. Harrison's system involved a subsidiary spring.)

The most notable of Harrison's horological achievements was the design and construction of a really accurate watch which would maintain its performance at sea under storm conditions and so enable a ship's longitude to be easily calculated. The problem was that no early clocks would keep going at sea, let alone remain accurate. Several large prizes were offered from 1598 onwards by crowned heads, governments and scientific assemblies for such a timepiece—an all-weather ship's chronometer. In 1659 Christiaan Huyghens constructed a sea-going pendulum clock, but it did not survive bad weather. The British Government, in 1714, offered a reward of £20,000. John Harrison devoted much of his life to the task of meeting their stringent demands, producing various spring-powered chronometers. His large and cumbersome Number 1 chronometer (reported to have weighed seventy pounds) was fairly successful, and his Number 4 chronometer—essentially a large pocket-watch—eventually won him the full £20,000, although he had to petition George III to obtain his money, in 1773. It should be mentioned that his success was due to the improved quality of spring steel as much as to his own watchmaking abilities.

Several other accurate sea-going chronometers were developed during the eighteenth century by English and French clockmakers.

Night Clocks

The clockmaker's ingenuity has often been called upon to produce a clock which would enable the time to be known at night—without the owner being required to strike a light. From at least the fourteenth century up to the late nineteenth, night clocks were designed either with a light inside them or else to give the time by feel or sound.

One of the earliest known is a German alarm clock with touch-knobs, dating from about 1400. It has a sixteen-hour dial to allow for long winter nights, with a knob at each hour, and was almost certainly made for the use of monks. (There are braille watches today which can be read by touch—although with dials of twelve hours only.)

From the seventeenth century, clocks were made incorporating a candle or oil-lamp. These operated in different fashions, by shining through numerical cut-outs on a revolving hour-band, by shining through a transparent dial, or even by projecting the whole dial onto a wall in the manner of a slide show.

A fifteenth-century Italian monastic alarm clock.

A sixteenth-century German table alarm clock.

While some clockmakers tried hard to make their night clocks silent so as not to rouse the sleeper, others installed repeating chimes. The compromise is to be found in those clocks which would not make a sound until one pulled their cord, but would then chime the exact time in hours, quarters and minutes.

Ornamental Clocks

Although all clock- and watchmakers liked to achieve greater accuracy all the while, it was not the design and construction of precision timepieces which most frequently occupied them and brought their profits, but the making of luxury clocks for the rich—for royalty in particular. Under the patronage of three successive King Louis'—XIV, XV and XVI—French clockmaking prospered. Favoured clockmakers became members of the royal household and were commissioned to produce ever more spectacular timepieces in that frivolous and extravagant development of the baroque style known as rococo, with which the king could amuse and delight his court and guests. Basically these were either cartel clocks (wall-hanging) or mantel clocks (to be placed on a mantlepiece or table), although some were so large and heavy as to be floor-standing: the longcase clock was never popular in France, although a few were made. Many of these royal clocks were decorated with the gaudy gilding of ormolu work; others were inlaid with brass, tortoiseshell, mother-of-pearl or precious stones. The dials were frequently enamelled. The incorporation of deliberately asymmetrical statuary and exotic automata was common, with a variety of animals included as clock-carriers. One piece in resoundingly poor taste was that made for Louis XIV, in which miniature models of the crowned heads of Europe bowed to a model of the French king before striking the hours and quarters with canes.

This was also the period when repeater clocks and watches were in vogue. Made possible by the invention of the rack-and-snail striking method in 1676, these timepieces were designed to give the exact time by chiming at regular intervals. For example, a quarter repeater would, at each quarter, strike the hour and then, in a different tone, strike the quarters. A minute repeater would strike the hours, quarters and minutes *every minute*, using three distinct tones.

Elsewhere than in France there was no less restrained a demand for exotic timepieces by those who could afford them. In the 1760s a watch so tiny that it could be worn as a ring was made for George III by the English clockmaker John Arnold. He was paid £500, but refused £1,000 to make an exact copy of it for Catherine

The thirteen-inch dial of a very fine clock by the famous Quaker clockmaker, Thomas Ogden of Halifax. This dates from about 1750 and is believed to be the earliest known provincial example of a world time-dial clock.

the Great of Russia. But Catherine did buy the famous Peacock clock by another English maker, James Cox, at about the same time. This is a complex example of automation, with (among other delights) a mechanical peacock in a tree and a cricket which moves to indicate the seconds. Made in England at about the same time (third quarter of the eighteenth century) for a Chinese customer was an eight-day musical clock with mechanical lotus blossoms, which open and close as the hour strikes, and revolving glass rods in the base which give the impression of a waterfall.

By the end of the eighteenth century the domestic clock was no longer the luxury it had been. The mantel clock and even the longcase clock were being produced in plain styles for the lower end of the market. Yet when the first cheap, mass-produced clocks appeared early in the nineteenth century they were not English, not even Bavarian, but American.

American Clocks

The story of clockmaking in the USA is one of astonishingly rapid progress. A few English clockmakers emigrated to the New England colonies during the seventeenth century, but none of their work from that time seems to have survived. Indeed, few American clocks are known from before 1750, although clockmaking was an established trade, particularly in Boston, Newport, New York and Philadelphia.

In about 1803, Eli Terry of Connecticut began to mass-produce thirty-hour clocks, working on the system of interchangeability of parts rather than concentrating on making one more-or-less unique clock at a time, as had always been done up to then. At first his movements were made entirely of wood; later he used brass. Clock factories were set up and many hundreds of cheap clocks were exported, particularly to England.

It must not be supposed that all American clocks were of the cheaper variety. Some fine clocks of high quality and original style were produced, notably by David Rittenhouse and Simon Willard. An innovation by the latter was the banjo clock, for wall mounting, which became popular after 1800.

Scientific Clocks

During the nineteenth century there were frequent improvements in clock design. A leading inventor was Louis Breguet, Swiss but working in Paris. Between about 1780 and 1823 he produced many important inventions, several concerned with maintaining the accuracy of a watch when it is placed in different positions. Effectively he made the shockproof watch possible, and several of his ideas have been developed during the nineteenth and twentieth centuries.

Among many new escapements designed during the nineteenth century, one specialized type was the Double Three-legged Gravity escapement, for use in the great clock at Westminster (wrongly called Big Ben, which is the name of the bell only). This was developed in 1852 by Lord Grimthorpe.

A typical mid-nineteenth century single-fusee skeleton clock with pierced chapter ring and passing strike of one blow at each hour. The power for running and striking comes from the single-coiled spring housed in the cylindrical case positioned in the lower centre area.

(*Left*) Bain's Patent Electric clock, an early example of the electric clocks now taken for granted in most public places and offices as well as in many homes.

(*Right*) The mechanism of the Shortt synchronizer, the most accurate mechanical timekeeper yet devised.

All but the very cheapest clocks had, by this time, become scientific instruments, capable of very great accuracy. It was accuracy which most impressed the Victorians, not merely surface decoration. As measuring instruments in general became more exact, so finer tolerances could be achieved in clock and watch manufacture. Styles of clock design changed; from the 1830s clocks beneath glass domes became fashionable, particularly with their movements exposed. Other clocks were made to resemble aspects of the industrial revolution, such as railway engines and bridges. At this time the Bavarians and Swiss began making carved wooden cases in the form of chalets.

In 1880 Greenwich Mean Time became the standard time base for the whole of the United Kingdom, and four years later, after much international discussion, the Greenwich meridian was accepted by general international agreement as the prime meridian from which international time zones and degrees of longitude are calculated.

To demonstrate that clockmaking in the late nineteenth century was not entirely concerned with greater scientific accuracy, one needs only to mention the name of Carl Fabergé, goldsmith, jeweller and maker of exquisite *objets d'art* for the members of the Russian imperial family. From the 1870s until the Revolution of 1917 quite a number of superb clocks were made by him or under his supervision. Not only were these pieces richly ornamented with precious metals and jewels, they were also made to the very highest standards of accuracy in terms of timekeeping.

The next step in clock design was radical but slow to come into general use because of practical difficulties—the electric clock. The first working electric clocks had been built during the 1840s by Alexander Bain, a Scotsman living in London, but he was not wholly successful due to the relatively poor quality of the electrical materials available at the time.

A reliable electric master-clock system was produced in 1894 by Frank Hope-Jones and George Boswell. It had a pendulum which was kept in motion by a gravity arm powered by an electric battery. With only a small increase in voltage many secondary clock dials could be kept at the same time. The system is used to power networks of clocks in large complexes, such as railway stations and factories. Normally the hands are advanced only every minute or half-minute. Timekeeping is very good; the impulse is given to the pendulum symmetrically and in mid-swing, not upsetting it in any way.

The Shortt Free Pendulum clock makes use of this principle. The most accurate mechanical timekeeper ever made, it was invented by William H. Shortt in 1921, and was used in observatories until the quartz crystal clock was developed. Its accuracy is to within a few thousandths of a second per day.

The use of a mains current to operate clocks was mooted in 1895 but did not become practicable until 1918 in the USA and 1927 in Britain, when a reliable AC current, running at fifty cycles per second (cycles per second are more properly known as Hertz,

A pair of mains electric alarm clocks by Westclox.

abbreviated to Hz) in Britain and sixty Hz in the USA, was introduced. Since then millions of synchronous electric clocks have been made, although these are, strictly speaking, not clocks at all because they include no time-measuring device. Really, they are no more than frequency meters, completely dependent upon the mains supply. Inside the clock is an electric motor with a rotor which spins at the same frequency as the mains supply. This frequency is reduced by worm gears to the rotation frequencies required by the different hands of the clock. Power stations attempt to ensure that the AC frequency is maintained. Peak period demand for electricity may lead to voltage reductions, but these will be metered and made up at slack periods so that mains electric clocks do not build up a cumulative error.

Transistorized battery electric clocks have become popular over the last fifteen years. Working from a 1½-volt battery which lasts for about twelve months, they are generally cheaper to buy than mains electric clocks, although less reliable. Most are spring-driven, the spring being rewound electrically every few minutes. In others the battery drives a pendulum or balance. Solid-state electronic battery watches have appeared in recent years.

During the first half of the twentieth century there was a search for the greatest possible accuracy of time measurement for scientific (particularly astronomical) purposes. Different types of natural oscillation were investigated to see if any were sufficiently unvarying to act as accurate pacemakers, and this led first to the development of the quartz crystal clock and later to the atomic clock.

Crystals of quartz can be made to vibrate mechanically by the application of an AC current. This vibration is extremely stable in its frequency, and the oscillations can be matched by an electrical circuit and then applied to a synchronous motor attached to an electric clock. The timekeeping element of a quartz clock is a

A GPO technician watching the silvering of a quartz ring for a quartz primary frequency standard for use in, say, an astronomical observatory. Made to the highest degree of accuracy such standards are subjected to prolonged testing before despatch. The quartz is cut and ground to size and shape, and coated with silver to provide an electrode. In this picture we see the quartz standard under the glass vacuum dome; also under the dome is a piece of silver suspended between two terminals. Carefully controlled application of current provides heat to melt the silver, which sprays evenly over the quartz.

A simple atomic clock.

The third-generation caesium beam machine used by the National Physical Laboratory to evaluate ever more accurately the exact length of the second. This machine is nearly ten times more accurate than the second-generation machine, seen on the right.

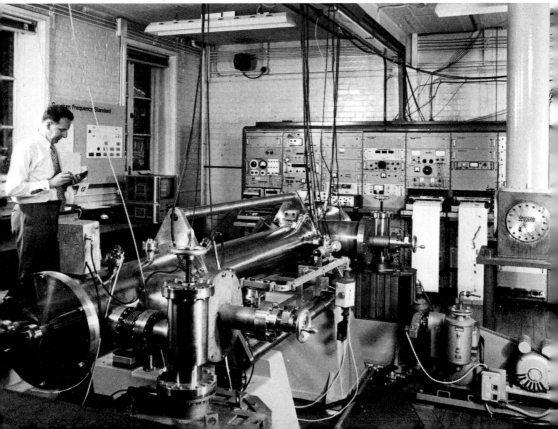

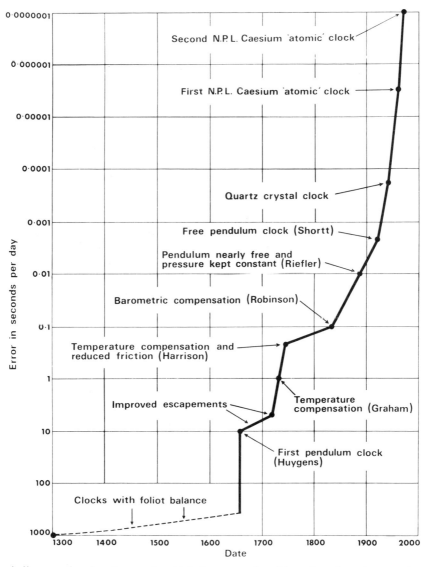

A diagram showing the way in which the accuracies of timepieces have improved over the centuries.

quartz ring about 2½ inches in diameter. The system is similar to a mains electric clock, except that the frequency of oscillations is far higher—about 100,000Hz. Consequently the gearing-down must be greater, but so is the accuracy. A laboratory quartz-crystal clock, kept at a uniform temperature, will remain accurate to within a few ten-thousandths of a second per day, and even a portable model can be relied upon to within one-fiftieth of a second per day. The first clocks of this type were developed by

Dr Warren A. Marrison in 1929. Digital watches, now becoming ever more popular, also rely on quartz crystals.

Even more accurate—indeed, the ultimate in timekeeping accuracy—is the atomic clock. This is a system which involves tuning a clock mechanism to the unvarying frequency of oscillations within atoms of particular types. The measurement of an atomic oscillation frequency was first achieved by Dr L. Essen and colleagues at the National Physical Laboratory, Teddington, near London, in 1955. A beam of caesium atoms was used (although later experiments were done with atoms of hydrogen, rubidium and thallium), being handled in a vacuum chamber by means of an alternating magnetic field. The oscillation frequency of caesium atoms has been measured as 9,192,631,770Hz ± 20. This is a level of accuracy reflected in the atomic clock operating at this frequency, and is equivalent to an error of just one second in 30,000 years. Accuracy of this magnitude is more than one can comprehend.

This oscillation frequency of caesium has more recently (1967) provided the sole definition of the second, and the basis of our entire system of timekeeping. The standard definition of length units is now similarly derived, although based on the length rather than frequency of waves or oscillations (but wavelength can be calculated from Hz by a simple formula). Thus one metre is defined as 1,650,763.73 vacuum wavelengths of the orange radiation emitted by atoms of the element krypton—a natural standard which can be reproduced to one part in a hundred million.

Our remote ancestors, noting time from the length of a shadow, would never have believed that before the year AD2000 Man would have constructed timekeeping devices millions of times more accurate than the apparent motion of the Sun itself.

CM

4 Bodytime

For modern man, time is a clock on the wall or an announcement on the radio or television. Suddenly we're late or early, or it's time to get up or to go to bed. Successive revolutions in the design of our chronometers have permitted finer and finer division of the day, into hours, minutes, seconds and even less until time is no longer a thing that any of *us* can measure as well as our instruments can. Our calendars are also accurate, guaranteed; the Moon rises on time, the Sun sets when they permit it. Time *consists* of these things—doesn't it?

It is my aim to show you in the following pages just a little of what *else* time is—how it can pervade and dictate and explain the behaviour and structure of all living things, from the simplest to the most complex, from the humblest protozoan to that most unhumble of all living creatures, Man. In short, a little of the complexities of bodytime—the time inside.

Biorhythms

In our naked environment, stripped of its concrete and glass, the dominant rhythm is apparently that of the Sun, circling through the heavens once every twenty-four hours. It gives us our day, our diurnal cycle, and has done since life began some three thousand million years ago.

Almost every physiological function that fluctuates in our bodies does so in time to this rhythm. Even such apparently insensitive organisms as plants show patterns of slow movement during the day which help them get maximum illumination from the Sun, while many genera (e.g., *Mimosa*) show a different movement during the night that helps protect their leaves from the cold.

This is, of course, nothing extraordinary, for the Sun as a giver of light and warmth is obviously ideally suited as our giver of time. What is extraordinary is that these diurnal rhythms are not dependent on the Sun or on any other environmental cycle; instead, we

"Even such apparently insensitive organisms as plants show patterns of slow movement during the day . . ." Here is a sprig of *Mimosa pudica* with its leaflets open (*above*) and closed (*below*).

follow a truly biological day. This has been proven quite convincingly over the last fifty years by research that was to create a considerable furore and produce new insights into many aspects of biology. The beginnings of this revolution were, however, humble enough, concerning as they did the subtle night movements of plants.

De Mairan, a Frenchman, was the first to record the observation

(1729) that certain species of plant closed their leaves at night and opened them for the day; even when kept in constant darkness, this daily opening and closing of the leaves was maintained. Duhamel went on to show (1758) that this cycle continued even when the nightly drop in temperature was prevented by constant heating. Thus, even in the absence of light or temperature clues, the plant appeared to know when it *should* be day, when it *should* be night.

De Candolle extended these observations (1832) by trying the effect of different lighting schemes; constant light caused the cycles to be maintained, but run a little fast—the 24-hour cycle became nearer to 22 hours. When he arranged for the plants to be lighted at night and kept in the dark during the day, after a period of adjustment the plants' cycle of movement was found to follow the artificial day and forsake the real one. Thus, de Candolle's plants *remembered* the normal cycle of day and night but would follow another cycle if subjected to it.

Despite this type of evidence for rhythms in plants, until fifty years ago the idea that they could keep time in any way resembling a clock was looked upon as preposterous. But, in the 1920s, Garner and Allard, two physiologists working for the US Department of Agriculture made a crucial discovery that was to completely alter our appreciation of biological time.

They had been asked to solve the problem of why a new strain of tobacco would not flower until so late in the season that the plants died from frost before they formed seeds. After examining the effects of the intensity and spectral composition of light, problems of transplantation shock, temperature and nutrition, they turned— almost in despair—to the effect of day length, and found the answer. By the simple expedient of keeping the plants in the dark except for ten hours of sunlight a day, they could make them flower in July instead of in the Autumn. The flowering response was triggered by reduction of the daylight hours to a critical length, the *photoperiod*. It has since been shown that of all the signs that plants might use to judge the time of year, this is the most reliable.

This was perhaps the first rigorously scientific demonstration that living organisms could accurately recognize a time interval, and use it to regulate their behaviour. These findings were soon put to good use in horticulture throughout the world, but their significance for biologists was not to be realised for a full ten years.

Just *how* plants measure this critical period with such accuracy was the subject of Erwin Bünning's famous and controversial hypothesis published in 1936. Stated simply, his proposition was that plants use an endogenous timing mechanism consisting of two phases of about twelve hours each that together constitute a circadian ("about a day" long) rhythm. In the "day" phase (the light-loving one) light falling on the plant encourages flowering, while in the "night" phase (the dark-loving one) light *inhibits* flowering. In formulating this hypothesis, Bünning had drawn on several other recently established facts—that bees could be trained to visit a feeding station at a certain time of day regardless of environmental changes, and that the timing of leaf movements could be drastically shifted by a single short exposure to light during a dark period.

This idea was in sharp contrast to the one prevalent at the time, that the duration of some light-driven biochemical reaction was the measuring device. Bünning's assertion that plants *timed* external events with an internal circadian cycle was considered by many as just too fantastic. There followed, however, such an explosion of evidence for the existence of a clock in animals as diverse as migrating birds, the honeybee, single-celled plants and cockroaches that the phenomenon of the internal circadian clock is now quite firmly accepted by most workers.

The existence of a clock in birds was inferred by Gustav Kramer in the 1950s from their extraordinary navigational abilities. Long-distance migration is one of the most startling feats of many bird species. The New Zealand bronzed cuckoo, for example, each year flies over 3,000 kilometres from its homeland *via* Australia to its tiny wintering grounds in the Solomon Islands. Most of this incredible journey is over featureless sea: how can the birds fly it so accurately, with no landscape to guide them? The newly born young then fly home *alone*, never having seen the route before; their parents precede them by at least one month.

Kramer chose to study caged European starlings for convenience, taking the direction of the nervous "flight fidgets" that appear during the migration season as an indication of which direction they would choose to fly in if free to migrate. The direction of these movements was always correct for the season, even if all the landscape and most of the sky were blotted out. Even when given just the image of the Sun to guide them, they oriented correctly by it—but if its image was shifted (by mirrors) then so too was their sense of direction: the birds appeared to navigate by the Sun. But how could this be when the position of the Sun changed hour by hour during the day?

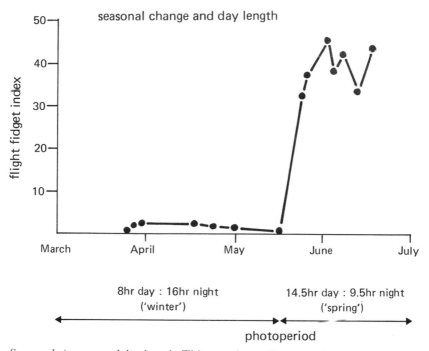

seasonal change and day length

flight fidget index

50
40
30
20
10

March | April | May | June | July

8hr day : 16hr night
('winter')

14.5hr day : 9.5hr night
('spring')

photoperiod

Seasonal changes and day length. This experiment illustrates the crucial dependence of seasonal changes—in this case, the appearance of migratory "flight fidgets" in caged bramblings—on day length. When the birds were kept in conditions of an eight-hour "day" through the migratory season no "fidgets" developed. After the "day" had been adjusted to 14.5 hours, the "fidgets" were soon observed.

To check these observations, Kramer trained the starlings to feed in the open at 7-8am in the east, so that they learnt to fly towards the Sun to feed; when released at 5.45pm, however, while the Sun was in the west, they still flew to the east—they corrected for the apparent diurnal movement of the Sun. Similarly, he trained birds to feed in the west all day in the open, and then kept them under cover with only a fixed image of the Sun to guide them. They then fed in the east at 6am, the north at noon, and the west at 5pm! That is, they applied their usual "correction" to the position of the Sun for the time of day even though the image of the Sun was stationary. In this last experiment, any possibility that the vertical angle of the Sun above the horizon was itself giving clues to the progression of time was ruled out.

Thus Kramer's starlings steered by a Sun compass, which they corrected for the time of day using some internal time-sense. The living clock was now fully and undeniably established; no longer were there any alternative explanations; his birds possessed an

Migration of birds can be a spectacular sight.

accurate internal chronometer. With this, a whole new avenue of research was opened up, and the new science of "biorhythms" was born.

Further proof of the internal nature of the clock came from similar work done at about the same time by Karl von Frisch on the Sun compass of the bees. In these creatures, he discovered that the location of a food source is communicated by the discoverer to the rest of the hive by a dance it performs on a vertical wall of the hive: this is in the form of a circle, with a diameter danced across it at intervals. The "up" direction corresponds to the Sun's position, while the angle that the diameter makes to this gives the bearing from the Sun that the other bees must take to find food.

That alone was quite remarkable but, in its dance, the bee also corrects for the change in position of the Sun throughout the day, without actually seeing this change.

Von Frisch then went on to investigate earlier claims that bees could memorize the time of day that food would be available. At first he thought that they might be responding to some clue from the environment, but he found that they arrived on time consistently even when constant light, humidity, temperature and atmospheric electric charge were maintained. None of these, therefore, could be providing a time clue. He finally decided that the clock was either dependent on some subtle, deep-penetrating form of radiation (as the timing persisted even deep in a salt mine) or was truly internal.

After many delays, he was able to put this to the test by training bees to feed at a fixed time in a room in Paris, and then sending them by aircraft to New York. If the clock was internal, the bees would emerge to feed in the new location precisely 24 hours after their last feed; if it was dependent on some environmental cue, then they would re-emerge after 29 hours, because New York's environment was five hours "behind" the Paris one. The bees emerged exactly on time by the Paris clock—their timekeepers were *internal* and independent of the environment.

This is not to say that the body clock is unaffected by the environment, for this would surely be a matter of "putting the cart before the horse": the internal clock must function to *aid* the organism's adjustment to the surroundings and its cyclic changes, not the opposite! Just like our own mechanical clocks, the accuracy of body clocks is maintained only by repeated reference to more accurate time sources.

The most important time reference for most organisms is the light/dark cycle of day and night, as was shown by Patricia DeCoursey in her work with flying squirrels. These creatures normally become active shortly after dusk and rest during the day, but when kept in constant darkness in her laboratory their usual 24-hour cycle of activity shifted to a period of between 24 hours 21 minutes and 22 hours 58 minutes. Each animal was remarkably consistent, with its own rhythm varying by only a few minutes a day. Thus, in the absence of environmental time clues, some clocks ran fast, and some slow.

But how did these light cues work? She found that in continuous darkness a single ten-minute pulse of light was sufficient to "reset" the clock, but only if it came just before the flying squirrel was about to resume its usually nocturnal activity. This pulse had two effects—it delayed the onset of the squirrel's activity, and also set the clock back by an equivalent amount.

In the wild, the timing of light sensitivity to just this period means that, even though a squirrel's circadian clock might run a little fast and tell him to become active while it was still light, the environmental stimulus—the presence of light—would be dominant, resetting the clock and delaying activity until it was safe.

Just how powerful a synchronizer light can be has been shown by subjecting plants to artificial days of varying lengths, such as 10 hours each of light and dark. The plants follow even so artificial a cycle, echoing it with a 20-hour circadian rhythm of their own movements. But there *is* a limit, and rhythms outside 20-30 hours are produced only under very extraordinary conditions (such as some forms of mental disturbance in monkeys). In particular, the more complex organisms seem to be more resistant to such manipulations: their rhythms are both more stable and more nearly 24 hours in length.

The Clock Inside

Compared to work on the *existence* of circadian rhythms, very little has been done on the mechanisms of the clock that produces them, or its location.

In *Acetabularia*, a very large single-celled alga, the rhythm of photosynthesis is maintained even if the nucleus is removed; yet, if the nucleus is replaced by another, the cell now follows the rhythm of the new nucleus. Although the nucleus seems to dictate the rhythm, in its absence the cytoplasm can still remember it.

In multicellular plants there is little evidence for a single centralized clock; cell cultures maintain a pronounced circadian rhythm (in the absence of any presumptive "master clock"), and the loss of circadian rhythms seen in constant conditions is quite rapid. In some plants (e.g., chicory), this loss of rhythm is actually due to a breakdown of synchronization, firstly between plants, then between flower clusters, and then between flowers themselves—the rhythmic movements of each flower are maintained until last. These are exactly the results you would expect if each flower had a separate clock, normally kept in step with the others by signals from the environment.

In multicellular animals some form of master clock would seem necessary to ensure efficient coordination of the activities of the various tissues, probably by means of hormonal or nervous messages. Yet some cell cultures retain circadian cycles of secretory activity, growth rate and metabolism in the absence of any such master clock, and it is possible to alter the phase and period in different organs of an animal quite independently. Most probably, both types of clock are present, and together determine the temporal behaviour of the individual cells.

In the 1960s Janet Harker performed a series of very delicate experiments on the cockroach that seem to show the clock's

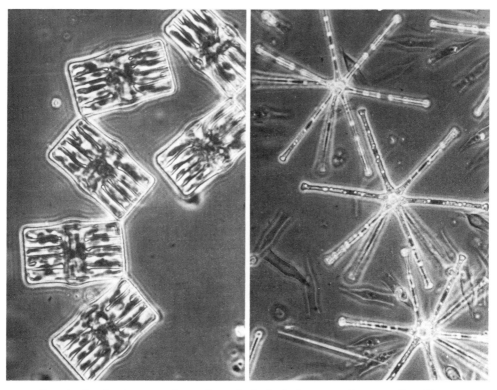

Even in the simplest organisms, algae, the bodyclock seems to have an effect. Here are examples of (*left*) *Tabellaria* and (*right*) *Asterionella*.

The cockroach was the subject of Janet Harker's experiments in the 1960s. Here we see *Blatta orientalis* on a piece of bread.

location—at least in this insect. Cockroaches are nocturnal creatures, normally having a period of running activity commencing
immediately after dark. She found that, kept in constant light, the
cockroach's rhythm was maintained for a few weeks, but then its
activity became completely random. In a series of very delicate
operations she found that removal of endocrine glands had no
effect on this cycle of activity, but that the suboesophageal
ganglion—a group of cells the size of a pinhead—was essential. She
finally showed that only four cells were important: with these particular cells destroyed by a high frequency cauterizer, the running
activity of the cockroach again became completely arhythmic.

Janet Harker carried out several more experiments, but her most
extraordinary result was obtained when she transplanted these
cells from a cockroach accustomed to one light-dark cycle into
another that was accustomed to a cycle exactly 12 hours out of step
with it. The resulting animal had two clocks set 12 hours apart.
The result was most striking and quite unexpected—the cockroaches rapidly developed intestinal cancer (usually very rare in
insects) and died.

Although it has since been shown that the timing of the cycle
also involves another organ (the corpora cardiaca), these results
represent the nearest anyone has yet come to locating the clock in
the intricately intermeshed control systems of a living organism.

The precise mechanism of the clock at the cellular level can still
be only a subject of speculation. Perhaps the best developed model
suggested so far is the "chronon" model of Charles Ehret.

To understand this model a little basic biochemistry is necessary. Cells are organized into two main compartments—the central
nucleus and the cytoplasm surrounding it. In the nucleus are the
chromosomes, complexes of the nucleic acid DNA (a huge double-
stranded molecule that carries the genetic information of the cell),
together with various proteins, while in the cytoplasm is a production line for the manufacture of proteins—the ribosomes. Proteins
are produced by first making copies of small segments of the DNA
(genes) in RNA (a nucleic acid somewhat less complex than DNA
and found throughout the cell), then transporting this mRNA to
the cytoplasm where it directs the synthesis by the ribosomes of the
protein coded for by each particular gene.

In Ehret's' model long sequences of DNA are organized into
"chronons". Copying of the DNA into RNA starts at the first gene
in this sequence, then halts until the protein corresponding to that
mRNA has been produced and has diffused back to "switch on"

the copying of the next gene (while also turning off copying of the first one). This process repeats itself many times until all the genes have been copied into mRNA in turn. In this way, the chronon functions like the escapement of a watch, measuring out many small intervals of time to span a much longer one. The final protein then restarts the whole process, after a considerable delay, a complete cycle of copying the chronon taking one day.

The process could be likened to the ticking away of the seconds by a one-day clock. After 86,400 ticks the clock tells us that 24 hours has passed, and needs to be rewound. Although we say that the clock has measured the passage of 24 hours, we could equally regard it as having counted a cycle of 86,400 ticks, which cycle happens to take 24 hours (unless the clock is slow or fast).

Ehret further proposed that there are many different chronons in the DNA of the cell, only a few of which would be running at any one time, depending upon the particular biochemical activity of that cell.

Although similar mechanisms to this are known to operate in the reproduction of the very simple viruses that infect bacteria, such as T4, MS2 and Q Beta, there is still little direct evidence for this type of cellular timing mechanism in higher organisms.

An alternative type of mechanism was suggested by H. Schweiger to take account of the fact that the rhythm of photosynthesis in *Acetabularia* is *not* stopped by chemicals that are known to inhibit the synthesis of RNA and proteins. He proposed that cellular timing could involve a simpler chemical oscillation, such as that observed in various multi-enzyme systems in cell-free extracts from yeast cells. These cyclic variations in the level of chemical intermediates occur over periods of only about five minutes, so that in order to produce a *daily* rhythm, some form of "counting" system or frequency reduction would also have to operate.

Whatever the mechanisms they depend upon, the existence of biological circadian rhythms is firmly established in very many organisms. Such cycles of metabolism and physiological function could have many rôles—perhaps the most universal of which is simply preparing the organism for the regular daily cycle of demands that will be made on it by the environment. In some species, however, a more specialized function has been shown to be the case.

Thus in migrating birds, and in bees, ants and spiders, the daily cycle is split into identifiable hours, so that these organisms possess a true clock by which they can tell the time of day. In many plants and birds of the temperate zones, on the other hand, the cycle functions instead as a stopwatch with which to time the length of the day and hence tell the time of year, thereby allowing the organism to anticipate the seasons.

Time is not just something that organisms respond *in*: they respond *to* it, derive information *from* it, and are structured *by* it. It is as much a dimension of their lives as any of the familiar spatial ones, and perhaps more so.

Time and Our Bodies

But what are the consequences for Man of these studies? How do *we* fare, having chosen to follow a mechanical clock and ignore our own internal ones? And, with our ability to control the environment, to choose when the day should begin and the night end, can we control our own internal clocks?

The existence of daily rhythms of body temperature was established over a century ago by Gierse (1842). The temperature maximum is a plateau from late morning to the early night, while the minimum (0.5–1°C lower) occurs in the early morning. These variations are due not just to the normal cycle of activity; they occur even if the subject is awake all night or confined to bed all day. Only in the very sick is the timing of the rhythm distorted significantly. Individuals do vary, however, in the rate at which their temperatures rise in the morning—perhaps explaining why some of us wake early and easily, while others still feel half asleep at 11am!

Excretion of urine also follows a circadian pattern, independent of activity: urine flow from the kidneys is at a minimum during the usual hours of sleep, and at a peak in the morning.

Numerous other circadian rhythms in body fluids and hormones have also been discovered: the coagulation time of blood is shortest at night, when the level of gamma globulins—molecules that confer immunity to infection—would be lowest (if results for the rat can be extended to the human); and sensitivity to a whole range of drugs varies cyclically throughout the day.

Each of these and many other rhythms begins at a different age: the cycle of urine flow at two to three weeks, body temperature at five to nine months, the excretion of various ions (reflecting muscle and nerve activity) at one and a half to two years, and the blood levels of adrenal hormones at about three years. Most of these rhythms persist in an only slightly altered form when the subjects are isolated from environmental cycles of light and dark and allowed to "free-run"—they are true endogenous circadian rhythms.

The importance of these daily cycles began to be realized only when Man started trying to adapt to unnatural cycles of activity. With the advent of World War II, we began to expect vigilance at any hour of the day or night—in factories or in the war zone itself. Since then, records of night-shift accidents have shown that the risks are greatly increased by such unusual demands, and assessment of mental and physical performance in the laboratory has suggested why, revealing a clear minimum in ability at about 2–4am that parallels the circadian drop in body temperature at this time.

Many of the crucial events in our lives—natural births, accidents and cardiac arrests—show a peak of incidence in the early hours of the morning, coinciding with our weakest physiological state and throwing an even greater strain on our ability to adapt.

Experiments where volunteers have been left to free-run in constant light show that the circadian rhythm of sleeping and waking is affected by light intensity—the stronger the light, the shorter the rhythm produced—but the effect is only minor. Conscious attempts to alter the length of the cycle have failed in most cases, although one individual has managed to lengthen his cycle in bright light from 19 to 25.6 hours.

Attempts to alter the sleep-wake cycle drastically to 48 hours by manipulation of lighting schedules initially seemed to have had more success. But, although the subjects managed to work a 34-hour day and then sleep for 14 hours, they reported feeling constantly drowsy, almost as if they were hibernating. Examination of their body-temperature cycles showed why—they had not adapted to the new work cycle at all. Thus, although we can will ourselves into unusual cycles of activity, we do not seem able to adjust physically, the adaptation being purely psychological.

While free-running, not all of the body's physiological rhythms need stay in their usual relationship. Thus one volunteer observed by Jürgen Aschoff showed a 23-hour cycle of sleep and calcium excretion, but a 25-hour cycle of water, sodium and hydrogen ion excretion so that the two cycles slipped out of their synchronization, except that, every 3 or 4 days, they came back into their normal relationship before moving out of step again. Later examination of the diary the volunteer had kept showed that, on the days when the two sets of cycles coincided, he had felt the most healthy. A similar dissociation of cycles is also often seen in sufferers from jetlag: could this be the actual cause of the symptoms of headaches, burning eyes, nausea, loss of appetite and sweating so familiar to some unfortunate travellers?

The effects of shift working, when people have to constantly readjust their sleep-wake cycles, have been examined in a survey of more than 1,000 industrial workers in the Rhône valley. 45% of the workers felt they could adjust to a seven-day rotation, but none of them had actually adjusted physically, as assessed by their rhythms of body temperature. Other studies have shown that shift workers are also more prone to stress-related diseases such as stomach ulcers. Proper adjustment to such an inverted activity cycle takes longer than a week, so that, for optimum physical performance, a permanent night shift is the best answer, although this poses obvious social problems.

In mature rodents, weekly inversion of the 24-hour cycle of light and darkness from the age of 58 weeks has been shown to produce a 6% reduction in average lifespan. This result is somewhat alarming when it is realized just how close this experience is to the normal routine of an airline pilot flying East-West routes (and hence constantly changing time zones). Jet fatigue has been taken quite seriously by most airlines, both as a cause of discomfort and as a possible safety risk. In tests conducted by the Federal Aviation Authority, aircrew that flew eastwards (USA to Rome) regained their earlier psychological performance in one day, but it took six days for their body temperatures to adapt, and eight days for the rhythm of the heart rate to adapt. Seasoned pilots often overcome the problems caused by crossing time zones by sticking doggedly to the time in their homebase: regardless of the local time in the country where they have just landed, they follow the clock that their body is used to.

In his advice to sufferers from jetlag, Hubertus Strughold suggested three methods of overcoming the problem: the traveller should arrive a few days early to allow time to adapt; he should adjust his own clock beforehand by gradually shifting his pattern of activity in time towards that of his destination; or he should perhaps ask for mild medication (stimulants or sleeping pills) to help in rapid adjustment. Erwin Bünning has suggested the possibility of readjusting the internal clock on arrival by short cycles of light and dark to cajole the internal clock into synchronizing with the new location's local time, while direct resetting of the body clock by means of hormones has been under investigation by at least one pharmaceutical company.

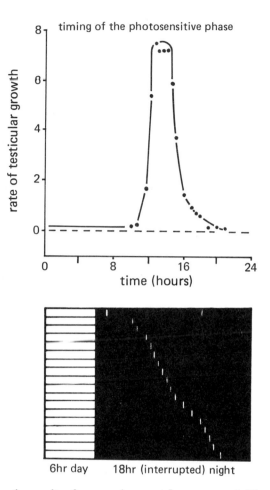

A graph showing the results of an experiment on Japanese quail. The quail were subjected to fifteen-minute periods of light at various times during their night, as shown in the lower part of the diagram; the graph above shows the rate of their testicular development, and demonstrates a marked peak around 12–14 hours after "dawn" corresponding to the photosensitive "switch" for this change.

Other Rhythms of Our Bodies

I have concentrated so far on the rhythmic phenomena that affect Man over the period of a day, for this is without doubt the most significant unit of time for the body. But cycles that run much faster and much slower than this are also known, although few have been investigated as thoroughly.

The two most obvious fast rhythms in our bodies are so fundamental that we rarely notice them at all. The regular beat of our hearts is with us all our lives, from only a few weeks after conception, while the slower cycle of respiration starts only moments after we are born. Both continue until the moment of our death.

The heart rate is normally controlled by nervous and hormonal factors that can vary the number of beats from 40 to 150 per minute in response to physiological and psychological stresses and demands. But, even if isolated from the body, the heart will continue to beat under the control of its intrinsic pacemaker system, the sinus node. The mechanism of this built-in rhythm is, however, not known.

Of the slow cycles in our bodies, the best known is, of course, the menstrual cycle of the female, which takes on average 28 days to complete—rather similar to the length of a lunar month (28.4 days) although there is little scientific reason to believe that the two are in any way related, as the menstrual cycle's length is notoriously variable, while of course that of the lunar cycle is not! Often associated with the menstrual cycle are (as 60% of women know to their cost) cycles of depression, irritation, headaches and decreased acuity. The existence in males of a similar cycle of more subtle hormonal and mood changes has also been suggested by at least one recent study.

As early as 1887 Wilhelm Fliess (a close friend of Sigmund Freud) suggested that both men and women were subject to a 23-day cycle of "maleness" (strength, endurance, courage) and a 28-day cycle of "femaleness" (sensitivity, intuition, love), which in combination determined their day-to-day mental and physical state. Although the original justification for this assertion now seems far-fetched, the underlying idea may be far less so.

Monthly cycles dictated by the Moon are seen in many sea creatures, as would be expected from the obvious importance to them of the tides that the Moon's waxing and waning cause. Far more surprising is the observation that recovery from operations in Man also seems to be affected by the lunar phase: thus the incidence of haemorrhages following throat operations has been reported as 82% greater during the second quarter of the Moon in Florida, while recovery from fractures of the head of the femur in Ireland has been reported as varying with the lunar phase.

Although annual cycles are quite apparent in birds (plumage, egg-laying, migration), plants (budding, flowering, dormancy), and many mammals (fur coloration, hibernation, reproduction), Man seems to show little evidence of a similar biological response to the time of year. It is true that there are regular annual fluctuations in statistics such as births and deaths, but these are quite adequately explained by social mechanisms: thus, in the summer, there are more accidental deaths because more of us are active, taking unusual risks on holidays and at weekends. Changes in hormone levels, such as those that produce seasonal behaviour in birds, do not seem to occur in Man, although the level of one thyroid hormone involved in temperature regulation does increase in the summer.

Cycles of illness, both physical and mental, are also known to exist and have puzzled some of the greatest medical researchers of the early part of this century. Variation of the electric charge of the atmosphere has been proposed as the cause of cycles of epilepsy and bronchitis by Svante Arrhenius (who won a Nobel Prize for his electrolytic theory of chemical dissociation), but, in general, the length of the cycle for any particular disease varies so much with the individual that the origins of the cycle are likely to be within the sufferer himself.

Periodic inflammation of the linings of the intestines and stomach (peritonitis) and periodic fluid retention (oedema) are recognized inheritable diseases, but it is quite likely that there are many other cyclic diseases not yet realized as such. Of the potentially cyclic psychoses, manic depression is perhaps the most easily recognized, with several cases of very stable 48-hour cycles having been reported. In one of these cases, when misled into a 22-hour cycle of activity, the patient's psychotic cycle rapidly became a 44-hour one despite most of his physiological rhythms staying on a 24-hour clock. Thus the manic depressive cycle seemed independent of his body rhythms.

Man would thus appear to be no more outside the rule of time than any other animal, and, although he may choose to ignore the natural rhythms of his body, a price must be paid.

Time and Our Minds

Man is, however, above all a *thinking* animal: any description of him must include as much psychology as biology. What then of the psychological aspects of time?

There seem to be three main aspects to Man's conscious appreciation of time: awareness of the time of day; perception of intervals of time; and extension of the conscious through time into past and future, by means of memory and anticipation.

The first aspect, a sense of the time of day, is perhaps the most obvious, as it is central to the structure of our highly organized civilization. Everywhere we look there are timetables and working hours and trains late and appointments missed. Life would be

". . . timetables and working hours and trains late and appointments missed . . ."
Here is Waterloo Station, London, in 1963 during a time when burned signalling
cables had paralysed rail traffic.

The London Stock Exchange on a morning in 1962. People such as businessmen
and lawyers, for whom the time of day is more important than it is for most of us,
are said to be able to estimate it more accurately.

Time and the athlete: the last few instants of the 5,000 metres at the Mexico Olympics, 1968: Gammoudi of Tunisia wins with Keino of Kenya in second place.

unimaginable without awareness of time—except perhaps during our summer holidays while lying in the Sun.

We rely heavily on our mechanical or electronic clocks. If you are late for an appointment, it would hardly do to say that you *felt* you were on time, but if your watch had stopped then who could blame you? Yet there are people who have an instinctive idea of the time without any need of watches, who can tell the hour within ten or fifteen minutes. Are they simply acutely sensitive to time clues in the environment, the chiming of distant clocks or perhaps the height of the Sun or Moon? The answer would seem to be no, for the similar ability to awaken from sleep at a predetermined hour is not affected by sleeping in carefully soundproofed and darkened rooms. (Only about 7% of the population can consistently awaken themselves in this manner, although most of us have probably experienced it at some time.)

A sense of the time of day seems to be at least partly learnt, for people such as lawyers and businessmen, who have a constant need to know the time, can estimate the time of day more accurately than those for whom the hour is of no particular importance. Perhaps it is merely a matter of learning how to recognize the subtle clues from our own body clock, in the same way that many animals have always done. Having delegated the responsibility for keeping time to mechanical timepieces, we seem to have lost the knack.

The second aspect, dealing with the perception of duration, has been the subject of many studies, such as those of the famous French psychologist Piaget.

The ability to estimate the length of an event in terms of standard "clock seconds" first develops in children at the age of six to seven years, and is well established by eight years. Young children tend to overestimate the length of any period; indeed, for them time always seems to drag. In the language of body rhythms, their "psychological clock" could be said to run fast.

Although well developed in adults, the accuracy of estimation of time intervals can be upset by many factors, giving us further insight into the mechanisms involved. Perhaps the most crucial finding is the effect of varying body temperature, through either disease or direct experimental manipulation. A high temperature (such as experienced during a fever) causes the subject to count faster than he would normally, showing that his clock is running fast. A similar result is found with bees: when trained to feed at the same time each day, they will arrive late if kept cold overnight, or early if kept warm.

Expressed mathematically, the effect of temperature on the estimation of time intervals is very similar to its effect on the rate of a simple chemical reaction. This suggests that the mechanism we use to measure short intervals involves a quite simple metabolic clock—unlike that involved in the circadian rhythms, which is accurately corrected for temperature.

This variation of time estimation with body temperature may be one reason why children always complain of time dragging—their bodies are warmer than ours, due to a higher metabolic rate. Similarly, in old age, body temperature is lower and time appears to move faster. Almost certainly, however, other factors are involved.

If a subject is asked to judge the length of an interval during which he is presented with information (e.g. in the form of a film) an interesting fact emerges. His estimate of the length of the time interval increases with the level of complexity of the information presented and with the amount that he remembers of it. Thus your mind's impression of the length of any period of time is partly dependent on how active it was during that time. A week on holiday filled with so many varied activities seems far longer than a week in the daily routine of work, while a baffling lecture will seem long because your mind was having to work hard trying to make sense of it all. This again may be a partial explanation of the child's impression of time dragging—each day is crammed with so much that is new and often incomprehensible; nothing is easily dismissed as mundane by the developing mind.

The young mind is also more easily misled by irrelevant information. When a five-year-old is asked to draw lines slowly and carefully for a short period, and then to draw lines as fast as he can for the same time, he always identifies the second period as the longer one, because of course there are more lines to show for his efforts!

As we get older we learn to compensate in part for such errors, but at any age, women always tend to be less accurate in their estimates, probably because they are more affected by context than men. The most important step in the development of accuracy is the emergence of the concept of an abstract unit of time with which to compare a duration. Without this a young child is incapable of ignoring the content of the interval he is trying to time; with it, an adult can at least try! Our estimates of short intervals will also vary predictably during the day, in a circadian rhythm that parallels the change in body temperature and frequency of the brain's alpha waves.

One of the few methods of improving our ability to estimate time intervals is hypnosis, perhaps because it smooths out some of the ripples of mental activity that usually upset our judgement.

Estimates of long intervals are particularly dependent upon clues from the environment. Subjects isolated from such clues—in caves or in other free-running experiments—frequently report having taken "a short nap" when, in fact, they've slept for a full eight hours. On emerging from isolation they may underestimate the length of their stay by as much as a half.

Our sense of time can be also distorted by chemical means, reminding us of the physiological origins of time perception so obvious in the simpler animals. Thus while barbiturates, nitrous oxide, or lack of oxygen slow down our rate of subjective time, drugs such as the amphetamines, LSD, or thyroxine speed up our own clocks, and make us overestimate the length of time intervals. It seems to be a general rule that drugs which accelerate our metabolism cause our internal clocks to run fast. Even such mild stimulants as the caffeine in the tea and coffee we drink have a measurable effect. One striking result from such studies is that the extent of the time contraction or dilation produced by chemicals is not the same for all activities. Psilocybin (a drug related to LSD) produces as much as an eight-fold increase in the frequency of the rapid eye movements involved in image fixation, but only an approximately two-fold increase in the rate of finger-tapping which the subject was asked to keep constant. This suggests that different activities may be associated with different time-givers; behaviour that involves more conscious control seems to be less easily dislodged in time.

There are occasions, however, when our mental processes seem to operate completely outside the normal flow of time. These events take place quite regularly and are an obscure but essential part of our lives. Such extraordinary events led Sigmund Freud to propose that "the processes of the Unconscious system are intemporal, that is they are not ordered in time, they are not modified by the passage of time, in fact they bear no relation to time". I am talking, of course about dreams—a subject Colin Wilson will deal with also, in Chapter 7.

When we can recall the content of a dream (and we normally recall only perhaps 1% of them), not only is it often completely disorganized in time (the dead attend their own funerals, effects precede causes, people are in two places at once), but a whole day or week somehow seems to telescope into the objective reality of only a few moments. Volunteers rendered unconscious for just a few seconds by drugs such as acetylcholine bromide nevertheless remember long and vivid dreams on awakening. Do we then, in our dreams, experience some vastly accelerated rate of time, once relieved of the inertia of our conscious physical bodies?

A similar phenomenon can occur in the mind of a man when suddenly faced with seemingly inevitable death. Hackneyed as it might seem, some of the stories of a dying man's "whole life flashing past before his eyes" seem to be well authenticated. The following account, told by a Swiss geologist named Albert Heim who fell while mountain climbing, is quite typical: "During the fall, a flood of impressions swept over me. What I thought and felt in those five to ten seconds cannot be told in ten times as many minutes. I watched the news of my death reach my loved ones and in my thoughts, I consoled them. Then, I saw as if on a distant stage, my whole past life playing itself out in numerous scenes . . ." Is this another example of some ultra-high-speed world, confined by the walls of our skulls?

The answer comes from careful examination of the *actual* content of dreams rather than our interpretations of them. Thus, a young

German physiologist named Sturt recalls a dream he had when awakened by his father ringing a large bell twice. He dreamt that he was giving a demonstration dissection to his class: first, he rang a bell to call for the body to be dissected, then he dissected it, and finally rang again for the remains to be taken away. The actual duration of the dream was fixed by the time it took his father to ring the bell twice—five seconds at most. Yet the dream seemed to him to last about an hour! However, when Sturt tried to recall what actually happened during the class, he couldn't remember a thing. His dream had consisted of only two or three images—the body coming in, his being in a demonstration class, and the remains being taken away. Using these few static images provided by the "intemporal subconscious" and experience of the everyday situation to which they related, his conscious mind had *constructed* a conscious temporal equivalent, by placing the events in sequence and then adding in sufficient time to make the whole fit its experience.

Such a conscious "trick" cannot, however, explain such extraordinary feats of perception as Mozart's being able to write down both parts of the *Misère* of Gregoria Allegri after having heard it through only once. Nor can it even begin to explain the abilities of such "human computers" as the Brahman girl, Shakuntala Divi, to extract the twentieth root of 42-digit numbers or perform even more complex calculations without any particular deliberation or training. Experience of phenomena such as these, coupled with the demonstration of unusually rapid learning under hypnosis, suggests that the speed at which we learn, perceive and even think is probably restricted more by our expectations of ourselves and our attitudes towards time than by any inherent physiological limitations. We will be returning to these points in Chapter 7.

The third aspect in the development of an appreciation of time is the temporal extension of the conscious from the present into past and future. At the age of two years, a child probably has memories stretching as much as a month into the past, but is only beginning to *anticipate* as much as a day ahead. As his interest in the past extends to events beyond his own birth (age eight years), he begins to see the patterns of time and use them to anticipate the future— by four years he recognizes the seasons, by five he is more precise and thinks in terms of particular days: Christmas or his birthday.

As we get older, not only does our memory extend further, but we gradually become more adept at using it. Thus, although a child may remember the temporal order of similar events (school holidays, birthdays), when asked to put two *different* sets of events into order he cannot because they are not remembered together. As adults, we have learnt to do this, but by a process of logic—in our minds, the two series of different events are still stored separately.

In the young the major preoccupation is with the future, while in the old, it seems to be the past: "when I grow up . . ." gradually becomes "when I was young . . .". In between, the adult mind must extend into both past and future, drawing on its memories to predict and plan ahead, a true Janus.

Lifespan

Ageing is both the most obvious and the most inevitable mark left by time on the living body. Although it occurs in all many-celled animals and plants, ageing is not, however, universal—many (but not all) single-celled organisms can grow and divide indefinitely under the right conditions.

As they grow older multicellular organisms pass from an initial stage of growth to one of maturity, where the processes of deterioration are largely balanced by repair activity, to the stage of senescence, where deterioration gets the upper hand. As they deteriorate they become more vulnerable to stresses such as infection, shortage of food and extremes of temperature, until finally one of these is sufficient to kill them.

Different species age at different rates, so that each animal has a typical average lifespan. For the vertebrates, as a general rule, the larger the animal the longer its life expectancy. This initially suggested that the metabolic rate determined the rate of ageing, as the smaller the animal the faster its metabolic processes. Furthermore, techniques that increased the metabolic rate (such as raised environmental temperature in cold-blooded creatures) also decreased the lifespan. It seemed as if all organisms had the same length of life, but some chose to live it faster.

However, this theory did not fit the primates very well—Man, for instance, should live only thirty years, judging by his size. The relationship seems more complex, with the brain size perhaps involved as well: a larger brain would mean better control of metabolism and fewer metabolic "accidents"; also, putting the chicken before the egg, a longer life makes a larger brain worthwhile (to store the extra information accumulated).

But why should animals age at all? If the body is capable of repairing itself throughout maturity, why does it not continue maintaining itself forever, and so become immortal?

152

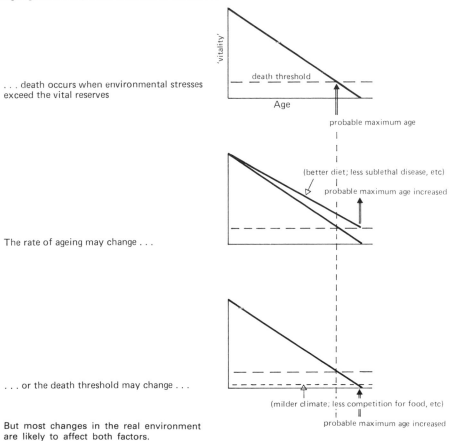

Ageing involves a gradual reduction in vitality . . .

. . . death occurs when environmental stresses exceed the vital reserves

'vitality'

death threshold

Age

probable maximum age

(better diet; less sublethal disease, etc)

probable maximum age increased

The rate of ageing may change . . .

. . . or the death threshold may change . . .

(milder climate; less competition for food, etc)

probable maximum age increased

But most changes in the real environment are likely to affect both factors.

The answer is not obvious. It seems to be that the potential for immortality is irrelevant under natural conditions, because of the high mortality from causes other than ageing—i.e., predation, starvation, accidents, disease. Thus, there is little or no selection for long-lived species. Coupled to this is the definite advantage of deferring any deterioration to an older chronological age—if an animal becomes weak when young, before breeding, it is at an evolutionary disadvantage, but not if the weakness can be delayed until after it has mated and passed on its genes. (Ageing of individual cells has also been suggested as a "safety net" that limits the growth of rogue cells that might otherwise kill the organism—cancer cells have lost this limitation, are immortal in cell culture, and *do* kill the organism.) Yet a further disadvantage of immortality in a species is the inability of the individual organism to adapt fundamentally to a changing environment: on the small scale, as

153

with bacteria, this can be accommodated by having a fantastically rapid breeding rate coupled with a fantastically high ratio of loss from each generation owing to accident and "predators"; on the larger scale, where such breeding rates are not possible because of limitations on the rate of cell growth, an immortal species would rather rapidly die out, since either (a) colossal overpopulation would result in disaster, or (b) its members would be unable to cope with both the changing environment (leading to accidental deaths) and the depredations of shorter-lived but far more adaptable predators.

In the case of Man, of course, many of these mechanisms are now inverted: most deaths at least involve ageing changes, and an old man can influence the survival of his genes by helping his progeny. Although it is still rather early to say, selection for longevity in *Homo sapiens* is a definite likelihood in the future.

The mechanisms involved in ageing are not as yet clear, but several possible models have been described. These are of two main types—models where ageing is pre-programmed, and models where it is produced by the accumulation of random errors.

Pre-programmed senescence requires direct involvement of genetic information in the deterioration of cells; ageing is seen as a direct extension of the processes of cell differentiation started in the embryo. In Strehler's "codon restriction hypothesis", for instance, he postulates that when cells differentiate (i.e., change from an uncommitted "general purpose" type of cell to a specialized type, such as a white blood cell) they do so by losing the ability to process all but a few classes of genetic information. As a result, they lose the ability to repair many of the structures in the cell, since the information for replacing these can no longer be decoded. As these essential structures gradually wear out, the cells die.

Ageing due to the accumulation of errors has been postulated in many forms. One of the earliest models was Burnett's "somatic mutation hypothesis", based mainly on the evidence of radiation-induced life-shortening in animals. He proposed that background radiation produced a gradual accumulation of random changes in the genetic information of the body's cells sufficient to upset their functioning and cause deterioration. In this model, longer-lived

(*Above right*) This woman's beauty treatment combats ageing only in a superficial sense: she will live no longer because of it. But scientists are now researching ways of slowing the ageing process in a real sense.

(*Below right*) Some people seem already to have fought and defeated ageing successfully: this Turkish man is a hale and hearty 156 years old.

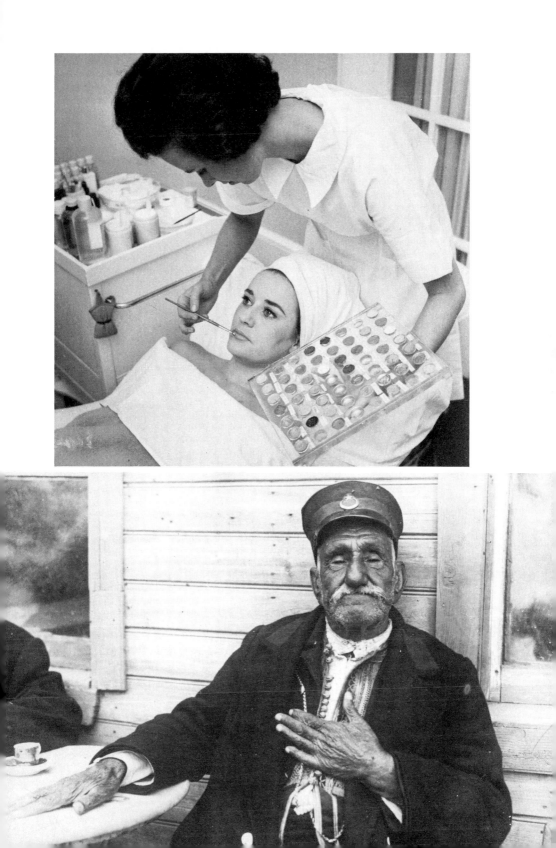

organisms owed their longevity to better repair mechanisms and multiple copies of the crucial genes: when one copy was spoiled, they could call on spares.

Whatever the mechanism of ageing, one of the most important questions for modern medicine—increasingly concerned with the care of the aged—must surely be: how can it be slowed down? Experiments with cell culture have shown that addition of vitamin E or hydrocortisone greatly prolongs survival of cells in the laboratory, suggesting that these agents might reduce the rate at which we age, but this promise has not been fulfilled.

In the 1930s, McCarthy showed that reducing the calorie intake of rats could greatly increase their lifespan by halting their development at the juvenile stage for long periods. When fed an adequate diet, they then matured and aged naturally. While such an extreme course is obviously not desirable in the human population, it shows that dietary manipulation may be a powerful tool, even if it is just to reduce the incidence of obesity and related disorders such as maturity-onset diabetes.

Obviously, much more research is needed into the phenomenon of ageing. Why is it that there are communities in the Andes where living to 140 years is deemed unremarkable, while there are children born to die before they are thirty years old from Hutchinson-Gilford syndrome, a form of premature ageing? We don't know yet—but we are trying to find out.

This, then, has been a brief look at time from the viewpoint of a biologist. Time not as a clock, but as an organism; not as a single rhythm, but as a whole symphony, so entwined with the processes of the living body as to be indistinguishable from life itself.

EP

5 Mutable Time

Time presents a curious dichotomy to the scientist. On the one hand, in many of his experiments he is concerned to measure how phenomena vary with time, and in so doing he measures intervals of time with great precision, taking for granted that time passes by at a uniform rate. On the other hand, when he investigates the properties of time he finds that its nature turns out to be very different from this tidy uniformity. The rate at which time passes depends upon the relative states of motion of different observers, and there are circumstances in which even the order in which events occur may be disputed. These mutable aspects of time are explored in the special and general theories of relativity, within which space and time are seen to be intimately related rather than being separate entities. The laws of nature do not appear to prohibit the possibility of time running backwards, and the reason why time should appear to us to flow uniquely in one direction is by no means obvious; indeed the whole concept of the "flow" of time seems to be highly unsatisfactory.

It is to these perplexing facets of time that this chapter is devoted.

Absolute Time and Space

"Absolute, true, and mathematical time, of itself, and from its own nature, flows equably without relation to anything external, and by another name is called duration." Thus in his *Philosophiae Naturalis Principia Mathematica*, published in 1687, did Isaac Newton encapsulate what has become the everyday "commonsense" attitude towards the nature of time. The notion of the relentless uniform flow of time is deeply rooted in modern western civilization, ruled as it is by the clock. Our lives are regulated by this mechanical contrivance—or its more modern electronic counterpart—and as we watch the seconds tick away we are acutely aware of the inexorable passage of our allotted span from

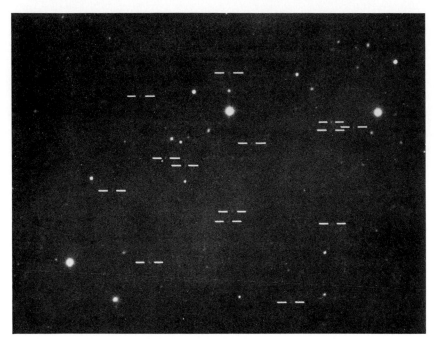

A field of very faint—i.e., very distant—galaxies in Coma Berenices. The wisps of the galaxies can just be seen between the pairs of white lines. They are among the most distant objects that can be seen, the light from them having taken about three billion years to reach us; that is, we are seeing them as they were at about the time that the first life appeared on Earth.

birth to inevitable death. In everyday experience the onward flow of time is self-evident, and there appears to be no reason to question its regular progress.

Newton went further in his discussion of absolute time: "All motions may be accelerated or retarded, but the flowing of absolute time is not liable to change. The duration or perseverance of the existence of things remains the same, whether the motions are swift or slow, or none at all." In other words, the state of motion of a body has no effect on the rate at which time passes, or upon the length of time for which that body may exist. Absolute time flows at an even pace throughout the Universe, and all observers, wherever they may be and however they may be moving, will agree on the times at which events occur and on the rate at which time flows. This accords with our everyday view of the world. An hour is an hour at every point on the Earth's surface or, for that matter, on the Moon; it is still an hour whether we are lounging at home in an armchair or driving a fast car. We shall see later that such self-evident notions do not appear to hold true in the Universe

Sir Isaac Newton (1642–1727), probably the most important figure in the history of science, whose notion of "absolute time" held sway until the late nineteenth century.

at large, even although they do seem to fit in with that very limited body of local experience which we call "commonsense".

As was noted in Chapter 1, the idea that time may be likened to a straight line (linear time) and that time flows uniformly in one direction so that "the future" becomes "the present" and then "the past", is a comparatively recent one. Most early civilizations had some kind of intuitive concept of "cyclic time", whereby the history of the Universe proceeded in a repetitive series of cycles. Regular periodic phenomena in nature were more obvious to earlier generations of mankind than to those of us who live in the man-made world of urbanized technological civilization. Day followed night in an inevitable cycle, while the changing phases of the Moon and the annual sequence of seasons were each clear evidence of a cyclical pattern. On a more sophisticated level, the pattern of lunar and solar eclipses is broadly repeated over a period of 18 years 11 days—the Saros cycle which was known to the Babylonians several thousand years ago. It seems very likely that the massive stone circles and alignments of megalithic Man were used for studying cyclic celestial phenomena and for the prediction of events such as eclipses. In particular, the Maya of central America constructed the most elaborate and accurate calendars based on natural periodicities, and in their view there was an over-riding cycle of 260 years during which major historical events would repeat themselves.

There were good reasons for regarding time as being cyclic in the sense that the Universe keeps repeating the same basic cycle of stages, or that the sequence of events is cyclic in time. As G.J. Whitrow,* and others, have discussed, this is a very different thing from suggesting that time itself is truly cyclic. If this were so, time would be closed like a loop. There could be no difference between the Universe going through a single cycle of events, and its going through a succession of identical cycles, since any distinction would imply that there exists some more fundamental non-cyclic time against which different cycles could be judged. Since the first and last events of a cycle coincide, they could not even be distinguished. Indeed, if time were closed like a loop, past events would also be future events!

The rise of Christianity brought a wholly different conception of time to the western world. God had created the world, and the history of the world was proceeding towards a definite end, the Last Judgement. Events would not repeat themselves; the Crucifixion and Resurrection of Christ were unique events. Time was finite and linear.

The analogy between time and a line, linear time, was developed notably by Galileo and by Isaac Barrow, who, in 1669, resigned the Lucasian Chair of Mathematics at Cambridge in favour of Newton. Like a line, time has only one dimension (length), and just as we can think of a line as a continuous entity or as a succession of points so we can consider time to be a succession of *instants* or the continuous flow of one instant. According to Barrow, ". . . whether things move or are still, whether we sleep or wake, time pursues the even tenour of its way"; again, "we evidently must regard time as passing with a steady flow". These ideas clearly had considerable influence on Newton and are embodied in his concept of absolute time existing in its own right and flowing at a uniform rate.

In Newton's view there also existed an *absolute space*, a fundamental background to the Universe against which, in principle, it should be possible to measure the absolute motion of a body from one "absolute place" to another. "Absolute space, in its own

* See, for example, G.J. Whitrow, *The Natural Philosophy of Time*, Thomas Nelson & Sons, 1961.

nature, without relation to anything external, remains always similar and immovable." The twin Newtonian ideas of absolute space and absolute time became accepted as fundamental aspects of the Universe until the advent of the theory of relativity at the beginning of the twentieth century. We shall examine later the dramatic transformation of our view of space and time which relativity brought about, but it is only fair to point out that even today Newton's view of space and time fits in with what most people would regard as a true description of the natural world.

The concept of the uniform flow of time has been greatly emphasized by the ever-increasing precision of clocks. Modern atomic clocks operate at accuracies equivalent to an error of one second in 150,000 years. There exist hydrogen maser clocks which, over limited periods, can maintain their timekeeping with a precision of better than one part in 10^{14} (one hundred million million); extending the previous analogy, such accuracy is equivalent to an error of one second in three million years! Atomic clocks enable us with ease to measure irregularities in the rotation period of the Earth, itself the former standard of timekeeping. Faced with worldwide standards of timekeeping of such high precision, the constant dependence on clocks and watches, and the temporal regulation of practically all aspects of our lives—eating, sleeping, working, playing—it is hardly surprising that we have been coerced into accepting without question the onward *flow* of time. We are very conscious of the existence of the present moment, that instant which we call *now*, and of its forward movement. The future flows towards the present and then flows on to become the past. We tend, with Newton, to envisage the entire Universe as having a present state of existence, the *universal now*; the future Universe has still to come into existence, the past Universe has passed out of existence. We feel sure it should be possible to consider the whole Universe as it exists now. Central to the Newtonian view of time is the concept of simultaneous events; if there is absolute time then events in the Universe are simultaneous if they occur at the same moment of absolute time, and there should be no reason for dispute between observers as to whether or not particular events were simultaneous.

There were, and are, powerful philosophical objections to the whole idea of flowing time. How fast is time flowing? The notion of flow implies motion with respect to time, and if time flows its rate must be measured against what—time itself, or some more fundamental kind of time? Without some external agency against which to make measurements how can we attach any meaning to Newton's assertion that absolute time "flows equably"? One could well argue that, since nothing can flow with respect to itself, then time cannot flow.

Among those of Newton's contemporaries who did not accept the concept of absolute space and time was Gottfried Leibniz. He maintained that neither space nor time has separate existence. Space is merely the separation between objects; it does not exist in its own right any more than "friendship" or "hatred" might be said to have separate existence ("friendship" implies that A is a friend of B, "hatred" implies A hates B, but friendship or hatred cannot exist in isolation). To talk about "space" is merely to talk about the relationship between objects, but space would not exist without objects. Likewise time is merely the *order of events*, not an entity in itself. We can derive the idea of time from the sequence of events in the Universe, but there is no absolute time consisting of a regular series of moments which exist in themselves.

Leibniz' view is made plain in his own words: "I hold space to be merely relative, as time is . . . I hold it to be an order of co-existences, as time is an order of successions." Or, again, ". . . instants, considered without the things, are nothing at all . . . they consist only in the successive order of things . . ." and, "I don't say that matter and space are the same thing. I only say, there is no space where there is no matter; and that space itself is not an absolute reality."

The view propounded by Leibniz is known as the *relational theory* of space and time and in a number of respects it accords more closely with the modern, relativistic concept than does Newton's theory of absolute space and time. It regards events as being more important than instants of time, and it suggests we may draw analogies between space and time. Within a somewhat different framework, the verbal battle between relationists and absolutists still continues today.

Be that as it may, it was the Newtonian view of time and space which found favour and which became the central dogma of science in the eighteenth and nineteenth centuries. By the late nineteenth century, however, physical science had run into a number of difficulties which stemmed from the acceptance of absolute space and time, and the resolution of these difficulties by Einstein's theory of relativity demolished these twin pillars of classical physics. The magnitude of the upheaval cannot be overrated. Time, space and commonsense could never be the same again.

The Emergence of the Theory of Relativity

Physics based upon Newton's laws of mechanics made great strides between the late seventeenth and the nineteenth century. Newton's laws of motion formed the cornerstone of classical physics; they were in essence as follows:

I Every body continues in a state of rest or of uniform motion in a straight line unless acted upon by a force.

II The rate of change of momentum of a body is directly proportional to the applied force.

III To every action there is an equal and opposite reaction.

The first law established that uniform motion in a straight line is the natural state of motion of bodies and that force is not required in order to maintain motion (as had previously been supposed by authorities such as Aristotle). Force, according to Newton, was only necessary in order to *change* the state of motion of a body: i.e., force produced acceleration. The second law takes the idea further, stating that if a force is applied to a body the resulting acceleration takes place in the direction of the force and that the magnitude of acceleration achieved depends upon the strength of the force *and* the mass of the body. The law is often written as: *Force = mass × acceleration.* "Mass" in this law defines the proportionality between the applied force and the resultant acceleration, and is referred to as the *inertial mass* of the body, inertia being the resistance of a body to a change of motion. It is due to our own inertia that we feel ourselves pressed back into our seats during rapid acceleration or, conversely, that in a car crash passengers are "hurled forward" from their seats: in reality they are continuing their uniform forward motion while the car containing them has halted abruptly.

Newton's law of universal gravitation was another key foundation in physics. It states that each particle of matter attracts every other one with a force which depends upon the masses involved and upon the inverse square of their separations. For two masses, m and M the strength of the force of attraction, F, acting on each body is given by $F = GmM/d^2$ where G is the gravitational constant and d is the distance between the masses. Mass in this context is referred to as *gravitational mass*, and, as we shall see later, it is a matter of great significance that the gravitational mass and inertial mass of an object have precisely the same value (at least within the limits of very accurate measurement). Gravitation was regarded in Newtonian theory as a force which acted instantaneously at a distance; i.e., the gravitational attraction of one

163

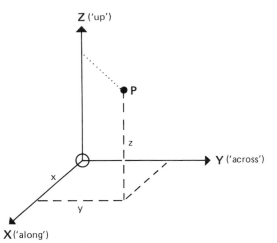

Fig. 1 A rectangular Cartesian frame of reference. The directions X, Y and Z are perpendicular to each other; these directions comprise the axes of the frame of reference. The position of point P relative to the origin O is given by the coordinates x, y, z since, to get from O to P, one would have to travel a distance x along the X-axis, a distance y along the Y-axis, and a distance z along the Z-axis.

body was immediately experienced by a distant body.

An observer moving at uniform velocity and not subject to any kind of acceleration is known as an *inertial* observer, and Newton attached particular significance to such idealized persons. If such an observer is a scientist he will wish to make measurements of the Universe around him, and to do this he requires a *frame of reference*, a yardstick with which to make his measurements. Space has three dimensions, and we think of solid objects in space as having length, breadth and height. We can specify position in space or the size of a solid body by making measurements in three mutually perpendicular directions ("along", "across", and "up", or "x", "y", and "z"). To specify the occurrence of events we need also a device for measuring time (a clock). Any observer can devise a grid, or frame of reference against which to make these measurements, and the frame associated with an inertial observer is called an *inertial frame of reference*.

Clearly Newton's first law should apply in any inertial frame of reference. If three cars are moving parallel to each other but at different speeds along three adjacent lanes of a highway, then the driver in the inside lane and the driver in the middle lane will both be overtaken by the car in the outside lane. Both drivers (who are inertial observers) will agree that the car in the outside lane is moving at uniform velocity (although, of course, each will measure a different relative velocity between the third car and himself). In fact, all of Newtonian mechanics holds true in *any* inertial frame, and the relative speed of two laboratories has no effect on mechanical experiments carried out inside these laboratories. For example, if I were to stand still in the corridor of a smoothly running high-speed train and allow a pebble to drop from my hand, it would fall straight to the floor, accelerating as it did so in precisely the same way that an identical pebble would do if I performed the same experiment in my living room.

Although Newton was convinced of the existence of absolute space through which bodies moved with absolute velocities, he recognized that in practice we were in a position to measure only relative velocities. A car moving along a highway at 100 kilometres per hour has that velocity relative to the surface of the Earth, but the Earth itself is spinning on its axis and moving around the Sun, while the Sun is moving around the centre of our star system (the Galaxy), and so on. However, he was in no doubt about the existence of absolute time and the possibility of all inertial observers agreeing on the times at which events took place. If one observer saw two events which he reckoned were simultaneous, all inertial observers would agree that they were indeed simultaneous. A particular instant of time was the same particular instant of time everywhere in the Universe.

In the nineteenth century great strides were made in the understanding of the motion of electrically charged particles moving under the influence of electric and magnetic forces. These developments were synthesized by the Scottish physicist James Clerk Maxwell, who suggested that each charged particle is surrounded by a *field*, an invisible aura which acted upon other charged particles which were placed within it; i.e., the field of one particle exerted a force upon the field of another particle. This concept was different from Newtonian gravitation, in which gravity was a force which acted instantly across the distance between one mass and another. An electrically charged particle, in Maxwell's view, was influenced by the field rather than by a force acting directly upon it from the other charge. One might draw a crude analogy by saying that Jack fell down the hill because he encountered the steep slope of the hill rather than because of a force acting upon him from the bottom of the hill.

The idea of the field led to the suggestion that all of space must be filled with an invisible fluid within which the field could be embedded. This fluid medium came to be known as the *ether*. The motion of charged particles should generate waves which would travel through the ether, just as sound waves travel through air or water waves through water. The speed of these waves depends upon the properties of air and water respectively; likewise the speed of *electromagnetic* waves should depend on the properties of space. The speed of these hypothetical waves proved to be exactly the same as the measured velocity of light.* The implication was clear. Light was a form of electromagnetic radiation which travelled through the ether in the form of waves. Maxwell's equations predicted the existence of waves of longer wavelength than visible light and, in due course, waves of this kind, radio waves, were generated in the laboratory by Heinrich Hertz. Today we are familiar with electromagnetic waves of all kinds of wavelength, ranging from less than a million-millionth of a metre (gamma rays) to metres or even kilometres in the case of radio waves.

That light might be a wave motion had of course been discussed earlier, but Maxwell gave the theory a firm mathematical basis. Since all known waves existed in a medium, such as air or water, it was only reasonable to suppose that space must be filled with a medium which carried light waves. Space could not be completely empty, for how could a wave exist if there were nothing for it to "wave"? The case for the "luminiferous" ether looked good.

It was then suggested that the ether might be interpreted as Newton's absolute space. If the ether were at rest and filled all space, it seemed reasonable to take the ether to be the absolute standard of rest in the Universe, and if light travelled at a constant speed through this medium then it should be possible to devise experiments to show how fast the Earth is moving through the ether and so establish the absolute motion of the Earth. The most famous of these experiments was the Michelson-Morley experiment, performed first in 1881 by Albert Michelson, and again by himself and Edward Morley with improved apparatus in 1887.

The principle of the experiment is illustrated by the following analogy. Imagine a race between two boats, each capable of precisely the same speed through the water, which takes place on a river which flows at a uniform pace. The river is one kilometre wide. Boat A has to cross the river to a point directly opposite and return to the starting position, while boat B has to go downstream to a point one kilometre along the bank, and then return to the start. Which boat will win the race? The answer is that A will do so every time. This is explained in detail in the caption to Fig. 2, but

---

* The velocity of light was first determined in 1675 by the Danish astronomer O. Roemer from observations of the times of eclipses of satellites of the planet Jupiter. The presently accepted value is about 300,000 kilometres per second.

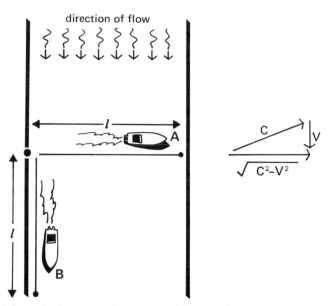

direction of flow

Fig. 2 A hypothetical powerboat race which A would always win. If we imagine that the river flows towards the bottom of the page with a current velocity of v, that both craft can maintain a steady velocity c, and that the width of the river is l, we are ready to start. A has to sail directly across the river and then back again while B has to sail downstream a distance l and return upstream to the start. As shown on the right, A has to aim slightly upstream to allow for the current and so, using Pythagoras' theorem, will travel across the river at a rate rather less than c; i.e., $\sqrt{(c^2-v^2)}$. A's time for the complete race will thus be $2l/\sqrt{(c^2-v^2)}$.

On the downstream leg B makes good time, travelling at a rate of $c + v$, since he has the current behind him; but on the way back he will be able to achieve a rate of only $c-v$: his travel time will thus be $l/(c+v) + l/(c-v)$; or $2lc/(c^2-v^2)$.

These mathematical expressions may look rather frightening, but a moment with paper and pencil will show that they derive from elementary schoolboy algebra. Taking a few moments more to divide A's time by B's time we arrive at $\sqrt{(1-v^2/c^2)}$.

But $\sqrt{(1-v^2/c^2)}$ must always be less than 1, unless $v = O$, i.e., unless the stream had no current; and so A will always win the race. We are, of course, assuming that c is greater than v; a fair assumption, since one can imagine the mayhem that would otherwise result as we attempted to stage our race!

the crucial point is that, although B completes the downstream half of his trip before A gets across (because he has the current in his favour), he more than loses this advantage on the return half (when he is moving against the current).

The argument regarding the ether went as follows: if the Earth is moving through the ether, and if light moves at a constant velocity through the ether, a ray of light sent in the direction of the Earth's motion and then back to its starting point should arrive later than a ray sent over an equal distance at right angles to the direction of the Earth's motion through the ether. The experimental apparatus

167

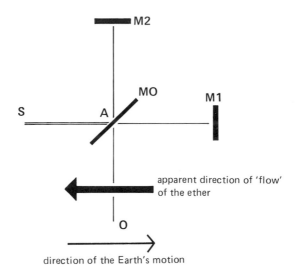

S ——————— A

M2

MO

M1

O

apparent direction of 'flow'
of the ether

direction of the Earth's motion

Fig. 3   The Michelson-Morley Experiment. This simplified diagram illustrates the basic principle of the experiment. A beam of light travels from source S to a half-silvered mirror M_0, which allows half of the beam to continue straight on to mirror M_1 while reflecting the other half to mirror M_2; both mirrors are the same distance from A. The reflected beams from M_1 and M_2 again meet M_0, and part of each travels to the observer O. If light were to move at a constant speed relative to the ether, and if the Earth were moving through the ether as shown, then the ether would appear to flow past the apparatus (rather like the flow of the stream in Fig. 2). That being the case, the beam reflected from M_1 should take longer than the other beam to complete its journey. The observer O should be able to measure the extent to which the two beams have got out of step. All attempts by Michelson and Morley to make this measurement failed, indicating that the postulated "flow" of the ether had no measurable effect on the two light beams.

is illustrated in Fig. 3. The Earth moves around the Sun at a speed of some 30 kilometres per second and, although that is very small compared with the velocity of light, the sensitivity of the apparatus was such that this sort of motion through the ether was well within its range. In practice, *no difference whatsoever* was measured between the travel-times of the two beams.

Since it was possible in principle (although rather unlikely) that at the time the measurements were made the Earth just happened to be stationary relative to the ether, the experiment was repeated at different times of the year, when the Earth was moving in different directions; it could not then be stationary on every occasion. Still no difference was detected. It gradually became apparent to physicists that *no* experiment could show the motion of the Earth through the ether. In terms of our analogy of boats on a river, the result seems to make no sense at all, for it seems to suggest that the speed of the river would make no difference to the time taken for the two boats to complete their journeys. They would both return at precisely the same instant.

When the results of such experiments had reluctantly been accepted some notable physicists attempted to preserve the notion of the all-pervading ether by rather devious means. Independently, in the 1890s, the Irish physicist George Fitzgerald and the Dutch physicist Hendrik Lorentz suggested that motion through the ether would affect measuring instruments by just the right amount to prevent motion through the ether from being detected. In particular, measuring rods (rulers, or whatever was used to measure length) would shrink in the direction of motion, and clocks would run slow. Any instrument designed to detect motion through the ether would thus fail to achieve its goal. Although this very convenient hypothesis in a sense "explained" the failure of the Michelson-Morley experiment, it also demonstrated—as the French mathematician Henri Poincaré pointed out—that the ether, if it existed, must always remain undetectable. If all experiments designed to detect motion through the ether were doomed to failure, there could be no proof that the ether existed. Something which, even if it exists, is wholly undetectable in principle as well as practice, is of no value to science. The ether turned out to be a wholly useless concept.

Special Relativity Arrives

In 1905 Albert Einstein swept away the creaking foundations of the classical view of space and time and resolved the problem of the Michelson-Morley experiment by means of his *special theory of relativity* which was published in that year. The theory was based upon two fundamental postulates.

The first of these was the *relativity principle* that "all inertial frames are fully equivalent for the performance of all physical experiments". This implies that, if a laboratory is moving at uniform velocity, the motion of the laboratory has no effect whatsoever on the outcome of any experiment carried out inside the laboratory. Einstein had been concerned that, although Newtonian *mechanics* was unaffected by the motion of inertial frames (i.e., mechanical experiments would give the same results inside laboratories no matter how fast the laboratories were moving), *electromagnetic* phenomena (such as the propagation of light) appeared to be based upon one particular frame of reference—the ether. It seemed to him that there was no good reason why one set of physical experiments should not be affected by uniform motion while

Albert Einstein (1879–1955), whose Special and General Theories of Relativity swept away the last vestiges of the hypothesis that time is absolute.

another set should. The relativity principle abolished that distinction and accounted for the failure of the Michelson-Morley experiment; the speed of the laboratory (the Earth) clearly had no effect on the experiment (the measurement of the time taken for rays of light to cover equal distances in different directions).

The second postulate was that light travels through a vacuum at a constant velocity in all inertial frames. In other words, the velocity of light measured by an observer is the same regardless of the relative velocity of the observer and the source of light. This appears to be nonsense, and it is an affront to what we call 'commonsense'. For. example, if two cars each travelling at 100km per hour suffer a head-on collision, surely the relative velocity of the impact is $100 + 100 = 200$km per hour. Special relativity suggests otherwise. If a spacecraft is approaching a source of light with a velocity of half the speed of light (i.e., 150,000km per second), what is the measured velocity of light as seen by the crew of that spacecraft? Commonsense suggests that, if light is travelling from the source at 300,000km per second and the spacecraft is approaching the source at 150,000km per second, then the relative velocity of light and the spacecraft will be 450,000km per second. According to special relativity, the velocity of this beam of light as measured by the crew will be precisely 300,000km per second. One and one does not necessarily make two, it seems, in the strange world of relativity!

This may seem absurd, but it is just what has been demonstrated by numerous experiments. Neither the speed of the source nor that of the observer has any effect on the measured speed of light. Are we to trust commonsense (which, after all, is based only upon everyday local experience) and reject the theory, or should we accept the results of carefully conducted experiments which show quite clearly that the Universe does not adhere to the simpleminded rules that we might wish to impose upon it? In view of the many successes of special relativity, we have no choice but to accept the latter. The fusion of the relativity principle and the constancy of the speed of light came in the special theory of relativity, and with its coming the whole idea of an ether, and the associated concept of absolute space, was discarded. The whole basis of more than two centuries of established physics was swept away at a stroke.

The Demise of Simultaneity

One immediate casualty of accepting the two postulates of special relativity is the concept of simultaneous events. If Newton's absolute time existed, events would be simultaneous if they occurred at the same instant of absolute time, and all observers would be able to agree that this was so; i.e., that the events happened at the same time. Special relativity decrees that two observers in relative motion will not necessarily agree that two events are simultaneous; indeed, it is most unlikely that they will so agree, unless the two events also occur at the same place.

How can an observer (you, me, or anyone else) decide which are simultaneous events? If the two events occur at the same time immediately adjacent to him there is no problem, but if the two events are separated in his frame of reference, then the situation is more difficult. If he is sitting at the mid-point of a long corridor, he can be sure that two events which take place at opposite ends of that corridor are simultaneous if light signals emitted from these events reach him at the same instant. That much seems self-evident.

Consider now a high-speed spacecraft passing a fixed observer (B) located on a space station. At the instant when he passes B, a crew member (A) at the mid-point of the spacecraft presses a

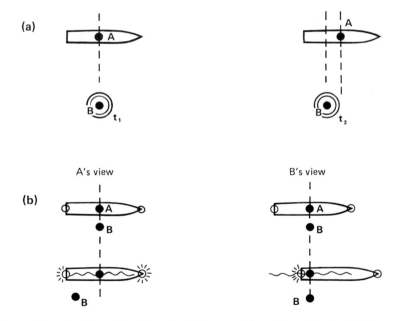

Fig. 4   The relativity of simultaneity. Observer A is at the midpoint of a fast-moving spacecraft while observer B is on a space station. At the instant in which A passes B he presses a switch which transmits two beams of light, one towards each end of his craft, the arrival of the beams causing lamps there to light up. In (*a*) A passes B at time t_1, and, in the time taken for the beams of light to reach the ends of the craft, the craft has moved forward to the position indicated (now time t_2). (*b*) From A's point of view, he is at rest within the spacecraft, and, since the signals have equal distances to travel, they will reach the opposite ends of the spacecraft simultaneously, i.e., from his point of view both lamps light up at the same time. But from B's point of view the light beams were emitted from a point in his vicinity (the point where A is) and will travel at constant velocity relative to him, covering equal distances in opposite directions in a given interval of time. Because the spacecraft has moved forward, the beam moving to the left has less distance to travel to meet the back of the craft than has the other beam to meet the front. Therefore B sees the rear lamp come on before the forward lamp: in his terms of reference the two events are *not* simultaneous.

switch which sends a beam of light to each end of the craft. The arrival of the beam (Fig. 4) causes a light to come on at each end of the spacecraft, and by the crewman's reckoning the two lights come on simultaneously. The "stationary" observer, B, has a rather different view. The first point is that, to satisfy the theory of relativity, the speed of light must be constant in *his* frame of reference. The second point is that, due to the finite speed of light, the spacecraft will have moved along by a measurable distance in the time taken for the crewman's signal to traverse the spacecraft; according to B, the rear of the craft has advanced some way towards the point at which A's light signal was emitted and the front of the craft has receded from this point. In B's view, the light signal has less distance to travel to meet the back of the craft than to reach the front. Therefore, he will see the light at the tail of the craft come on before the light at the nose. The events are *not* simultaneous from B's point of view although clearly they are so from A's viewpoint.

Who is correct, A or B? *Both* are correct within their own frames of reference, but neither is "more correct" than the other. There is no absolute truth about the matter, for there is no absolute time.

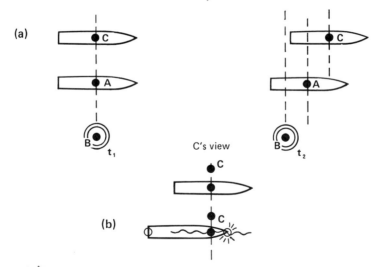

Fig. 5 A slightly more complicated example than that in Fig. 4 of the relativity of simultaneity. At the instant when A passes B a third observer, C, overtakes A in a faster-moving craft. The situation at some later time t_2 is shown in (a). In (b) we see the situation according to C. The light beams transmitted by A were emitted from a point in C's vicinity and must move at constant velocity in *his* frame of reference. Because his spacecraft is overtaking A's, C will conclude that the light beam moving to the right has the shorter distance to travel, and he will see the forward light come on first. He will agree with B that the lights did not come on simultaneously, but will insist on a *reverse* order of events.

Observers in relative motion cannot agree on which are simultaneous events.

We can go a stage further by imagining that the instant crewman A presses his button a second spacecraft overtakes him, heading in the same direction. Relative to the second craft (Fig. 5), and an onboard observer, C, the first craft is moving from right to left; i.e., relative to him, the first craft is moving backwards. C will note that A's light signal has less distance to travel to the nose of the first craft than to its tail. Therefore he will see the light at the nose come on before the light at the tail. Not only does C disagree with A over the simultaneous nature of these events, but he will see them happen in the opposite order to B's view. There are circumstances, then, in which inertial observers will disagree about the *order* in which events take place. As we shall see later, they *will* always agree on the order of events which are *causally* connected; i.e., where the first event causes the second to happen.

This situation is in flat contradiction to the Newtonian idea of "absolute, true, and mathematical time" which "flows equably without relation to anything external". With absolute time the order of events is uniquely fixed by the positions which they occupy *in* absolute time, and events should always be seen in their "correct" order. Special relativity demonstrates that time itself is relative, that no one estimate of the time at which an event occurred is more privileged than any other, and that—apart from the case of "cause-and-effect"—there is no "correct" order of events.

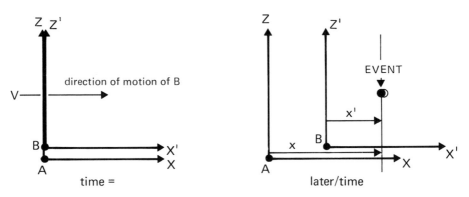

Fig. 6 Observers in uniform relative motion. At a particular instant B passes A, at uniform velocity v in the x-direction. At this instant A and B synchronize their clocks to read zero. At a later time (t according to A. t' according to B) an event is observed at position x according to AA, x' according to B. The relationship between the two observers' measurements is discussed in the text.

The Lorentz Transformation, and its Consequences

How are we to relate the observations made by observers in uniform relative motion? Imagine two such observers, A and B (again!), each equipped with a coordinate system whereby he can measure positions in space and a clock to measure time. Let us suppose that B passes close by A and at that instant they synchronize their clocks and agree that at the time of their encounter it was, say, "zero hours" (Fig. 6). After a time t has passed on A's clock, he sees an event occur at a distance x, measured according to his frame of reference. B in the meantime has been moving in the x-direction at a constant speed v. He too sees the event and assigns time t' and distance x' to it. According to Newtonian ideas, B's measurement of the time and position of this event will be related to the measurements which A made by the *Galilean transformation*, viz:

$x'=x - vt$: since B has been travelling for time t at velocity v in the x-direction, he will have covered a distance equal to vt, and will be that much closer to A;

$t'=t$: if time is absolute, both will agree on the instant at which the event took place.

Special relativity does not agree with the Galilean transformation. Instead, the measurements made by the two observers are related by a set of equations known as the *Lorentz transformation*. These equations are the ones devised by Lorentz to account for the undetectability of the ether, but they are a natural feature of special relativity. The (rather more complicated) relationships are as follows:

$$x' = (x-vt)/\sqrt{(1-v^2/c^2)}$$

where c denotes the velocity of light, and

$$t'=(t-vx/c^2)/\sqrt{(1-v^2/c^2)}.$$

Clearly the two times cannot be the same, unless $v = 0$.

There arise from the Lorentz transformation a number of effects which in everyday terms appear to be bizarre indeed, but which have been confirmed by experiment. These are: length contraction, time dilation, mass increase, and the concept of the speed of light being the greatest possible velocity at which a signal may travel, a velocity which cannot be attained by a material object. Our concern is with time in particular, but the other effects deserve mention too.

Length Contraction

The length of an object moving relative to one observer's frame of reference is less than the length of that object measured in a frame of reference in which it is stationary. If we take an object and measure its length when it is standing still, the value we get is called its *rest length*; if we measure the length of that same object when it is moving past us at high speed, we will measure a smaller value of length. In other words, moving objects shrink along the direction in which they are moving. They are shortened by the factor $\sqrt{(1 - v^2/c^2)}$, the Lorentz factor. A spacecraft moving past at 87% of the speed of light will appear to be only half rest length. What happens to the astronauts inside? From their point of view nothing has changed. From an outsider's point of view, their lengths will have contracted in equal proportion to the spacecraft itself, but there is no way that the inhabitants can measure or be aware of this contraction. The faster such a craft moved relative to us, the shorter it would become until, if it could achieve the speed of light, it would have no length at all.

In fact the situation is completely symmetric. If two spacecraft pass each other by, the crew of the first will note that the second craft is suffering length contraction, while the crew of the second craft will be equally convinced that it is the first which is contracted. Once again, special relativity emphasizes that there are no absolute standards of measurement in the Universe. Each crew is perfectly correct in its inference but neither can say that the other is "wrong".

Here, then, is hope for those who have an automobile which is too long to fit into the garage. If you drive in fast enough, at a large fraction of the speed of light, you can get any automobile into a small garage (provided that the garage has an infinitely strong wall to stop you at the end), since the automobile will have shrunk in length from the garage's point of view. You might well feel that the symmetry of the length contraction effect would pose a problem here. After all, from the driver's point of view the garage is approaching him, and *it* should appear to be shortened. This is true but, when the front of the automobile hits the back wall of the garage and stops dead, the back of the automobile will not yet "know" that this has happened. The back will continue to move forward until such time as the shock of impact is transmitted through the vehicle from the front to the back. Since the shock cannot travel faster than light, the back will always move far enough forward to get inside the door! If you are unconvinced, the

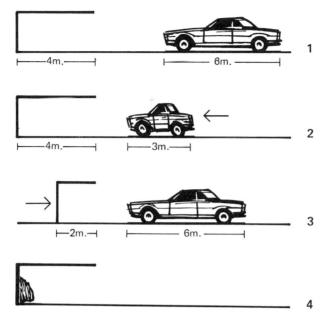

Fig. 7 The relativity of length contraction; how to make sure you can always get your large automobile into your small garage. (1) You are faced with getting your 6m-long automobile into your 4m garage. (2) Drive your automobile at 87% of the speed of light towards the garage. *In the frame of reference of the garage*, the automobile will shrink to half its original length because of the length contraction effect. (3) You will be alarmed. *In your frame of reference* the garage is approaching you at 87% of the speed of light and so will appear to be reduced in length to only 2m while your automobile is still 6m long. (4) But control those nerves. Assuming the back wall of your garage is infinitely strong, your automobile will crash into this wall and *will* fit into the garage after stopping. From the viewpoint of case (2) the back of your automobile will not stop until it receives the information that the front has stopped. This cannot be sooner than the time taken for light to travel the 3m from front to back; in this time, the back will travel a distance of 2.61m (i.e., 87% of 3m), and there is no doubt that the automobile will fit into the garage. And in the view of case (3) the back continues to travel forward for at least as long as the time taken for light to travel 6m, and so travels 5.22m to fit snugly into the garage. In both cases there may, of course, be a repair bill.

example is worked out in the caption to Fig. 7.

An interesting sidelight to this admittedly artificial example is the fact that in special relativity a perfectly rigid body cannot exist (the concept of a "rigid body" being another aspect of Newtonian mechanics). Any body which stops dead must become compressed, for the back of the object cannot "know" the front has stopped in a period of time less than the time taken for a signal to pass from the front to the back. No signal can travel faster than light, and light has a finite velocity. Therefore the front and back cannot stop at the same time, and the body cannot be absolutely rigid.

Mass Increase

The mass of a body at rest is called its *rest mass*. If a body is moving relative to an observer its measured mass (m) will be greater than its rest mass (m_0), and the closer the body approaches to the speed of light, the greater its mass becomes. The relationship is simply:

$$m = m_0/\sqrt{(1 - v^2/c^2)}.$$

At 87% of light speed, the mass of a moving object is double its rest mass, and as the velocity approaches closer and closer to that of light so the mass increases until, if a body could be made to travel at the velocity of light, its mass would become infinite. This is a disturbing factor for would-be interstellar travellers who, having such enormous distances to cover, would wish to travel as fast as possible. The faster the spacecraft moves, the greater its mass becomes and the greater the quantity of energy which must be supplied to increase its velocity by even a tiny amount. All the energy available in our entire Galaxy would be insufficient to accelerate even a single atom to *precisely* the speed of light; it could approach the speed of light very closely indeed, but it would never quite achieve it. Here we see in an obvious physical sense how travel at the speed of light is impossible for material objects, and how the speed of light represents an apparently impenetrable barrier to ultra-high-speed transportation. If a body cannot be made to travel at the speed of light, it seems obvious that it must be impossible for anything to travel faster than light. However, this may not be the whole story, and we shall return a little later to the possibility of faster-than-light particles.

Mass increase is a phenomenon which has been confirmed in laboratory experiments. Particle accelerators in nuclear laboratories are capable of boosting subatomic particles to very large fractions of the speed of light, and it is readily apparent from the results of such experiments that particle masses do increase in the proportion which special relativity predicts.

A directly related phenomenon which emerges from the theory is the equivalence of mass and energy: energy may be converted into mass and mass may be converted into energy. This is exactly what is happening with the increase in mass which we have just been discussing: the energy which we supply to a moving object in our endeavours to increase its velocity is partially absorbed into the increasing mass of that body. The relationship which Einstein derived must be one of the best known of all physical equations: where E represents energy, m represents mass, and c denotes the speed of light, then $E = mc^2$. That is, the energy obtained from the destruction of a certain amount of matter is equal to the mass destroyed multiplied by the square of the speed of light. Since the speed of light is a large number, and the speed of light squared a very much larger number, it follows that a large quantity of energy may be released from a small amount of matter. This relationship gave the clue to how the Sun and the other stars are shining. Inside the Sun the process of nuclear fusion is taking place. At a temperature of about fifteen million degrees (Centigrade) in the core of the Sun the lightest element, hydrogen, is being converted into the second lightest element, helium, and in the process a certain amount of matter is converted into energy. Every second the Sun destroys about four million tonnes of matter, and this sustains its colossal outpouring of radiation.

The equivalence of mass and energy, derived from what to most people seemed to be a wholly abstract theory of space and time, gave us the key to understanding the stars, to the generation of nuclear power and to the development of the nuclear bomb. The lesson to us is that no scientific theory, however divorced it may seem to be from everyday events, can be regarded as irrelevant: we cannot predict the practical consequences of a pure scientific development.

Time Dilation

An inertial observer will find that the rate at which time passes on an object moving relative to him is slower than the rate at which time passes in his own frame of reference. If he could observe a clock on board a fast-moving spacecraft he would see that the hands of this clock moved round the face of that clock more slowly than would the hands of a clock sitting on the table beside him. According to special relativity, there is no doubt about the matter; time passes more slowly on fast-moving objects than on "stationary" ones. From the Lorentz transformation we find that the time interval, $\triangle t$, between two events (e.g., two consecutive ticks of a clock) as measured by the stationary observer (A), and the time interval, t' measured by a moving observer, are linked by

$$\triangle t' = \triangle t \sqrt{(1 - v^2/c^2)}.$$

The observers will agree on the time interval only if the two observers are stationary relative to each other (i.e., if $v = 0$). For example, if B is moving at 87% of light speed so that

$$\sqrt{(1 - v^2/c^2)} = \tfrac{1}{2},$$

and if the clocks are synchronized to read $t = 0$ (and $t' = 0$) when B passes A, what time interval will be recorded on B's clock when 2 hours have been registered on A's clock? Our formula clearly tells us that B's clock will register $2 \times \frac{1}{2} = 1$ hour. B's clock is indeed running slow compared to A's clock.

Time dilation is a real effect, and it affects everything. Not just mechanical clocks, but atomic processes and all physical phenomena are affected in equal proportion. Above all, the biological clocks of the crew of a fast-moving spacecraft would be affected. The relativity principle is quite specific about this. The statement that "all inertial frames are fully equivalent for the performance of all physical experiments" implies that there is no way that phenomena taking place *inside* a closed laboratory (the spacecraft) can reveal the speed at which that laboratory is moving, provided that its speed is uniform. If the ageing processes of astronauts were not affected in the same proportion as the slowing of clocks then the crew would notice that they were growing old more rapidly relative to their (slow) clocks than they would do back on Earth. The relativity principle prohibits this possibility. If *any* process taking place within the confines of the spacecraft did not accord with the time dilation effect it would imply that there exists some kind of absolute time, and that there is such a thing as absolute space and absolute velocity. Many people when confronted by the phenomenon of time dilation, while being prepared to accept that perhaps mechanical clocks may run slowly at high speeds, refuse to accept that bodytime and the ageing process are likewise affected. But there is no avoiding the conclusion that human beings will be affected in precisely the same way as other material objects.

The closer a moving object approaches to the velocity of light, the more pronounced the time dilation effect becomes until, if it were possible to travel exactly at the speed of light, time would stand still, and any journey could be accomplished in zero time!

Each inertial observer has his own *proper time*. This is the time measured on a clock which he carries with him (and if he doesn't have a mechanical clock, it is the timescale at which natural phenomena and his bodily processes proceed). His proper time is the "correct" time as far as he is concerned, but the proper times of observers in uniform relative motion will not agree with each other. An inertial observer will note that *all* clocks moving relative to him run slow, so that no clock runs faster than a "proper clock". The time which an observer assigns to a distant event, based upon knowledge of the distance at which it took place, of the speed of the signal connecting the event to the observer (usually the speed of light) and of the proper time at which the event was seen to occur is known as *coordinate time*.

At ordinary everyday velocities the effect of time dilation is negligibly small. For example, a car travelling at 100 kilometres per hour will have v/c equal to about 0.0000001, and a clock in that car will run slow by only one part in 200,000,000,000,000; i.e., about one second in about seven million years. Even a rocket travelling at the Earth's escape velocity of 11 kilometres per second (about 40,000 kilometres per hour) will experience a slowing of its clocks by only about one second in fifty years. Put another way, an astronaut who travelled at this speed for fifty years would have aged by one second less than his Earth-bound contemporaries. At speeds which are a good fraction of the speed of light, the bonus in time to the would-be interstellar traveller would be considerable. At a steady speed of 87% of light speed, the journey time for the onboard astronauts would be half the time which that journey occupied in Earth time. A flight to a star system 10 light years away would take about 11.5 years by the Earth-based observer's clock, but to the crew aboard the starship the journey would occupy only half that time; i.e., 5.75 years (neglecting acceleration and deceleration times). At 99% of light speed the journey would be accomplished in just over 10 years of Earth time, but only 1.4 years would have elapsed on board ship. Given a high enough velocity any interstellar journey may be accomplished within the natural lifespan of the crew members. "Three score years and ten" could have very different significance for observers in rapid relative motion.

The most spectacular example of the theoretical possibilities for interstellar flight opened up by time dilation is given by considering the $1g$ starship. Because the long-term effects of the state of weightlessness on the human body are still unknown, it may be that long-stay astronauts (in space stations, for example) will require to generate "artificial gravity". This they can do by spinning their space station so that they experience a force which pushes them towards the rim of the station and gives a feeling just like weight. A spacecraft accelerates while its motors are firing, and this acceleration gives the astronauts a feeling of weight. If the spacecraft were to accelerate at precisely the same rate as a falling object near the Earth's surface (9.8 metres per second per second) then the astronauts would experience an apparent force (an "inertial" force) exactly equal to their normal weight on Earth. In fact this force would be quite indistinguishable from the sensation of weight. Ignoring the technical problems of fuel reserves and so on, if a spacecraft could maintain a constant acceleration of 1g, the most remarkable effects would occur. The crew aboard would feel a constant sensation of weight throughout the flight as their acceleration continued. As far as observers on Earth were concerned, the speed of the spacecraft would build up, getting ever closer to the speed of light but never quite getting there, and the time dilation effect would become greater and greater. After 100 years of Earth time only just over 5 years would have passed by on the ship, and it would have attained a range of about 99 light years, while 12 years of ship time would be sufficient to cover a distance of something like 100,000 light years (during this time just over 100,000 years would have passed on Earth), a distance equal to the diameter of our Galaxy. A complete circumnavigation of our Galaxy would occupy only about 25 years of ship time if the crew accelerated at 1g for the first half of the journey and decelerated at the same rate for the second half.

The time dilation effect has been checked experimentally in a number of different ways. One long-standing result concerns cosmic rays, charged atomic particles reaching the Earth from space, the origin of which is uncertain. These particles, striking the Earth's atmosphere, produce very short-lived particles called muons (or "mu-mesons") which disintegrate in an average period of about two millionths of a second as measured in a frame of reference in which they are at rest. In other words, the proper time between their formation and decay, which would be measured by a clock travelling with them, is two microseconds. Since these particles are produced some ten kilometres above the ground, even although they are travelling close to the speed of light, if there were no time dilation they would disintegrate long before they could reach the Earth's surface. At a speed of 300,000 kilometres per second, in two millionths of a second, they would travel only 0.6 kilometres. Because of their high speeds the time dilation effect is great enough (more than a factor of 10) that their lives are sufficiently extended to reach the ground, where they may be detected.*

This explanation was put forward in 1941 by B. Rossi and

---

* We can look at the situation from the point of view of the muon in a different way. At this very high speed length contraction will be so great that the distance which the muon "sees" from its point of formation to the ground is sufficiently short that it can get there within its own lifetime.

D.B. Hall, and since then laboratory tests on short-lived particles have produced similar results. For example, during an experiment carried out by Bailey *et al.* at the European Nuclear Research
Centre (CERN) in 1968, muons were maintained circulating in a
ring under the influence of a magnetic field at a speed in excess of
99.5% of the speed of light, such that the time dilation factor
should have been 12. In accordance with the theory's predictions,
the lifetime of these muons was found to be twelve times greater
than that of muons at rest.

The effect has been confirmed also by an experiment which is an
obvious one in principle, but which was impossible to achieve in
practice prior to the development of atomic clocks of outstanding
precision. In 1971, in an experiment carried out by J.C. Hafele and
R. Keating, four caesium atomic clocks were flown on regularly
scheduled commercial jet flights around the world (in easterly and
westerly directions to separate effects due to the clock's velocities
from effects due to the Earth's gravitational field) and the times
which they recorded were compared with the times registered on
fixed reference clocks maintained at the US Naval Observatory.
The results agreed with the predictions within an experimental
error of about 10%.

Time dilation undoubtedly exists, and there is no doubt that a
"stationary" observer (i.e., one at rest in his own inertial frame of
reference) would conclude that clocks on board a fast-moving
spacecraft were running slow compared to his proper clock. But
since no inertial observer is more privileged than any other, the
moving observer would be equally entitled to regard himself as
being "stationary" and to consider the first observer to be moving
relative to him. That being so he would regard the first observer's
clock as being the one which was running slow. The symmetry
between two observers in relative motion whereby each considers
that the other's clock is running slow leads us into a problem
known as the Twins Paradox, or the Clock Paradox, a source of the
utmost confusion, and the rock upon which many attempts to
understand relativity have foundered.

The problem may best be illustrated by considering an example
(one which may even become a practical example in the future).
Consider a pair of twins, John and Jane. Jane sets off on a voyage
to a distant star in a spacecraft which travels at a large fraction of
the speed of light, while brother John elects to stay at home on
Earth. John measures the time taken for the starship to reach its
target and to return to Earth. Due to the time dilation effect, time

The Seyfert galaxy NGC 4151. Some Seyfert galaxies are powerful sources of X-rays, and it has been suggested that this may be because they have at their hearts giant black holes, and that superheated material being dragged into these is emitting X-rays. Certainly, according to Einstein's General Theory of Relativity, the time dilation effects around such black holes, should they indeed exist, would be extreme.

passes more slowly on the spacecraft and the duration of the journey measured on board ship is less than the period of Earth time which has elapsed. Jane returns to find that her twin brother John is now many years older than she herself is.

This result is consistent with the phenomenon of time dilation, but the so-called paradox arises as follows. From what we said earlier, time dilation (and the other relativistic effects too) is completely symmetrical between observers in uniform relative motion. If John regards Jane's clock as running slow, Jane will regard John's clock as running slow. Consequently, as Jane speeds away from the Earth she is perfectly entitled to suppose that it is the Earth which is receding from her and to conclude that John's clock is running slow relative to hers. Likewise, on the return journey, she may regard the Earth as approaching her at high speed; again she will conclude that John's clock is running slow compared to her own. The apparently obvious conclusion is that when John and Jane again meet up, John will have aged less than Jane according to Jane's timescale, but Jane will also have aged less than John according to John's timescale. This is the paradox. How can Jane have aged more than John *and* less than John? Surely this is impossible. Does this mean that no difference arises in their ages after all, or—as some have suggested—that time "gained" on the outward journey is somehow "lost" on the return journey?

In fact the puzzle is easily resolved because the journey we have described is *not* completely symmetrical. Jane (in the spacecraft) is not entitled to suppose that it is the Earth which has receded from her and then returned, because a journey out and back involves *acceleration*. In order to return to the Earth it is necessary for the spacecraft to slow down, stop, turn round and accelerate once more to its high velocity.* Although velocity is not tangible, acceleration is. The inhabitants of a closed box *feel* the effects of acceleration; we are all familiar with this from the way we are thrown around in cars or aircraft as they accelerate or decelerate (turning a corner also amounts to acceleration as the direction of travel is changed). There is no doubt that the inhabitants of the spacecraft, Jane included, would be aware that it is they and not the inhabitants of the Earth who have been accelerated at the halfway point of the voyage. In her initial acceleration Jane changes her condition from being at rest on the Earth to a new frame of reference in which, because of length contraction, the distance to be covered to the target star is less than the distance measured in the frame of the Earth. Thus she completes her outward trip in a fraction of the time which John ascribes to the journey. When she halts at the target (as she is turning round) she is once again in a frame of reference in which the distance is consistent with John's measurement, but as soon as she accelerates to her cruise velocity for the return flight she is once more in a frame of reference in which the distance to travel is less. In the end there can be no doubt that it is astronaut Jane who has aged less than Earth-bound John.

If this does not seem wholly convincing, and if some suspicion lingers that such things would not happen in practice, perhaps a

---

* A spacecraft departing from the Earth would have to accelerate away from the Earth at the start of the voyage and decelerate again to achieve a landing on the return. But the paradox can equally be posed in terms of a "flying start"; i.e., a spacecraft passing the Earth at high speed and returning later to pass the Earth in the opposite direction at high speed.

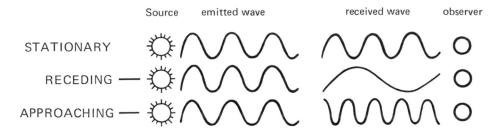

Fig. 8 The Doppler Effect. If a light source is stationary relative to observer O the light he receives will have the same wavelength as the light emitted from the source. If the source is receding fewer wavecrests per second will enter O's eye (i.e., the frequency of the wave will be reduced) and he will see a wave of longer wavelength than the emitted signal. Conversely, in the case of an approaching source, more wavecrests per second will enter his eye and he will see a wave of shorter wavelength than the original. A longer wavelength implies a reddening of light and a shorter one a "bluing"; hence the terms "redshift" and "blueshift".

specific example would help. Let us suppose that the distance to the star is 10 light years and that the speed of the starship is steady at 87% of light speed (we shall ignore time spent in acceleration and deceleration) so that the time dilation factor is such that time on the spacecraft passes at half the rate at which time passes here on Earth. We shall also suppose that John and Jane are equipped with the means to read each other's clocks. In practice this should not be too difficult as in each case the clock could transmit a radio or light pulse once every second, and onboard systems could count the arriving pulses to obtain the time registered on the other's clock.

As the spacecraft moves away from the Earth then—even in the absence of time dilation—Earth-bound John will find that Jane's clock is going slow. The time interval between successive pulses will be longer than one second of Earth time. The reason for this is quite simple; in the time interval between the emission of two successive pulses the distance between the spacecraft and the Earth will have increased so that the second pulse will have further to travel than the first and so will take longer to arrive, giving an interval of arrival between the two pulses which is greater than one second. This is an illustration of the well-known Doppler effect, whereby the number of wavecrests per second arriving from a receding source of light is less than the number of waves per second emitted from the source (a similar effect arises with sound waves in air whereby the pitch of an approaching sound is raised while the pitch of the same sound is lowered if it is receding). If the source is stationary (Fig. 8), the frequency of the arriving radiation is the same as that of the emitted radiation; if the source is receding, the frequency is reduced; and if the source is approaching, the frequency is increased.

John will note that Jane's clock appears to be going slow as the interval between successive pulses arriving from his clock is longer than one second. By the same token Jane will regard John's clock as running slow, since the frequency of arrival of pulses will be lowered. If we now add on the time dilation effect we will find that Jane's clock is seen to run even more slowly since at 87% of light speed it will run at half the rate of John's clock. Due to the symmetry of time dilation, Jane will make precisely the same observations of John's clock. Everything is symmetric so far.

Because of time dilation Jane will reach her target after 5.75 years of ship time, but in the Earth's frame of reference the journey will have occupied 11.5 years. If she immediately turns round she will take a further 5.75 years for the return flight and will arrive home after an elapsed period of ship time amounting to 11.5 years. From John's point of view the outward journey will take 11.5 years, but, since the target star is 10 light years away, he will not receive the signal emitted by Jane's clock as she reached the target until 10 years after that event; i.e. 21.5 years will have passed on Earth before John knows that Jane has reached her target. Since the return journey also occupies 11.5 years (making a total of 23 years in all) John must receive all the clock signals emitted by Jane's clock during the return flight in a period of only 1.5 years of Earth time.

During the outward journey, John receives 5.75 years' worth of Jane's time signals over a period of 21.5 years; i.e., Jane's clock is seen to run slow by a factor of 3.8. During the return flight John receives 5.75 years' worth of signals in only 1.5 years, Jane's clock therefore seeming to go fast by a factor of 3.8 (the Doppler effect more than compensates for the relativistic time dilation effect during this stage of the journey—Jane's clock appears to go fast rather than slow).

According to Jane, she gets to the target after 5.75 years and, because the observations made in the two inertial frames (spacecraft and Earth) must be symmetric, she will find that John's clock is running slow by the same factor of 3.8 (time dilation plus Doppler effect) during this stage of the journey. The time which Jane registers on John's clock at the instant when she reaches the target is 5.75 divided by 3.8; i.e., about 1.5 years. On her return journey, by symmetry, she must see John's clock speeded up by a factor of 3.8, so that in the 5.75 years of flight Jane registers the passage of 3.8 times 5.75 years; i.e., about 21.5 years on John's clock. Thus Jane has to agree that 23 years have passed on John's clock while only 11.5 years have passed on her own.

In essence, what has happened is that the situation between Jane and John remained symmetric *until* Jane turned round. Up to that, both could claim the other's clock was running slow. *Immediately* Jane turned round she would see John's clock running fast, because she was flying into the signals coming from it. John, on the other hand, would not notice Jane's clock speeding up until ten years after Jane had turned round. He would receive "speeded up" signals for only 1.5 years compared to picking up "slow" signals for 21.5 years. Jane receives equal periods of fast and slow signals from John while he receives unequal periods of fast and slow signals from her. The journey as a whole is clearly not symmetric. There is no paradox about the twins' experience, and about the fact that it is Jane who has reaped the benefit of time dilation.

The word "benefit" may be a little misleading here. True, Jane has gained in the sense that she has been able to accomplish a long spaceflight in half the time which has passed back on Earth, and has arrived back having aged half as much as her Earth-bound contemporaries, but she has experienced only 11.5 years of conscious existence. To take a more extreme example, suppose that the twins both have a lifespan of seventy years, but are separated at birth, Jane setting off at 87% light speed on the starship and returning to Earth at the end of her life, John staying on Earth. According to terrestrial observers Jane will have travelled in space for 140 years and will return 70 years after the death of John; but according to Jane's clock only 70 years will have passed, and she will have lived for 70 years only. Time dilation gives her the advantage of being able to accomplish journeys which might otherwise be impossible to achieve within her lifespan, but it does not give her a consciously perceived increase in the length of her life.

There are benefits, but there are drawbacks too. An astronaut travelling so close to the speed of light that the dilation factor was 100 would return to Earth after what, to him, was a ten-year journey to find that a thousand years had passed by on Earth and that he had landed in a world which, at the time he had set off on his voyage, lay a thousand years in the future. Society would have developed and changed perhaps out of all recognition during this period, and no trace would remain of the family and friends he had left behind; even his great-great-grandchildren would be dead and

The 200in (5.08m) reflector at Mount Palomar, one of the largest optical telescopes in the world. This is a view into the reflector, with an observer in the prime focus cage.

gone (barring some unforeseen development in longevity). He would have become a time traveller in a very real sense, but his time travel would be in one direction only—towards the future. He could *not* return to the past world of his erstwhile contemporaries. Time travel, then, is possible, but only in a forward direction; by making a two-way trip of sufficient duration at a sufficiently great speed it should be possible to arrive back on Earth at any future day of your choosing. The penalty to be paid is that this would be a journey with a one-way ticket. If the speed of light is a fundamental barrier, which it seems to be, and if the laws of cause-and-effect are to hold true in the Universe, journeys back in time must remain impossible. Only forward trips are permitted.

Of Tachyons, Time Travel, and the Order of Events

It has been suggested by a number of physicists that there may exist particles which *do* travel faster than light, and these hypothetical entities have been named tachyons. Now, special relativity shows quite clearly that it is impossible for a material body to travel at the speed of light, for its mass would become infinite. How, then, could a particle possibly travel *faster* than light? However, in one sense the mass increase formula of special relativity does not preclude the possibility of travelling faster than light, but only prohibits travelling *at* the speed of light. You may well argue that this amounts to the same thing; if you are driving along at a speed of 50 kilometres per hour and you wish to increase your speed to 70 kilometres per hour, then at some stage during your acceleration you must be travelling at 60 kilometres per hour. Surely, in a similar fashion, in order to be travelling faster than light, at some stage a particle must be travelling *at* the speed of light. However, what would be the situation regarding a particle which was travelling faster than light in the first place?

Take again the mass increase formula, $m = m_0/\sqrt{(1-v^2/c^2)}$. If we allow v to be greater than c, then on the bottom line we have the square root of a negative number, and this is what mathematicians call an *imaginary* number. If, however, we allow the rest mass also to be imaginary, then we find that, provided that v is greater than c, the hypothetical particle will have real mass and energy.* If the particle moves faster than light, it has a finite mass, and its mass reduces as its velocity increases. If we slow down a tachyon, its mass increases until, if its velocity were reduced exactly to the speed of light, then its mass would become infinite.† A tachyon, then, is a hypothetical particle which must always travel faster than light, while ordinary particles of matter (which may be called tardyons) must always travel slower than light. The speed of light remains a barrier which cannot be crossed by either tachyons or tardyons, but faster-than-light particles do seem to be at least a theoretical possibility.

There is as yet no experimental evidence of the existence of tachyons, although some anomalous results in cosmic-ray experiments reported in 1974 by R.W. Clay and P.C. Crouch of the

---

* An imaginary number is usually written down as a real number multiplied by the square root of minus 1; i.e., $\sqrt{-1}$. For example: $\sqrt{-4} = \sqrt{-1} \times \sqrt{4} = \sqrt{-1} \times 2$. $\sqrt{-1}$ is usually denoted by the symbol i so that $\sqrt{-4} = 2i$. We can now write the mass increase formula as $m = im_0/i\sqrt{(v^2c^2 - 1)}$, (where m_0 is the rest mass) so that, provided v is greater than c, the i on the top cancels the i on the bottom, and the mass is real.

† This implies that a tachyon must always have real mass. Although we have described it as having "imaginary" rest mass—and we cannot visualize what "imaginary mass" could be—since it cannot be made to come to rest, then we have no need to worry about its "rest mass".

University of Adelaide could possibly be interpreted in terms of tachyons. But if tachyons do exist and *if* they could be used to communicate information, then the most paradoxical results would ensue, and the whole ingrained notion that cause must precede effect would have to be abandoned. We saw earlier that there are circumstances in which observers in uniform relative motion will disagree about the order in which events occur; but, provided that no signal (carrying information) may be propagated faster than the speed of light, all observers will agree on the order in which *causally* related events take place. If event A causes event B then it must be possible for a signal travelling at or less than the speed of light to travel from A to B (otherwise A could not trigger the occurrence of event B). From the time dilation formula it can be shown that all observers in uniform relative motion will agree on the order of events. A specific example may help. A relativistic projectile is fired at 87% of light speed from a space station towards a target, which it destroys. The crew of the space station clearly see the explosion occur after the launching and make a note of the time interval between the two events. At the instant the missile was fired, a spacecraft passed the station moving in the same direction and at the same speed as the projectile. The crew of the spacecraft are, therefore, present at both events and because of time dilation will record a time interval between them of one half the time interval which the space station crew recorded (after making allowance for the time taken for light from the explosion to reach them). The two crews will not agree about the magnitude of the time interval, but they will agree that the explosion occurred *after* the firing of the projectile. Cause precedes effect in both frames of reference.

The order of events need not be preserved if a signal could be sent faster than light. If a superlight signal were sent from A to trigger an event at B, then a moving observer could well conclude that event B preceded event A; i.e., that effect preceded cause: all that would be required would be that the speed of the fast signal be greater than c^2/v, where v is the velocity of the moving observer relative to an observer in whose frame A precedes B. Furthermore, if a signal could be sent from point A to point B so as to arrive at point B *before* leaving A (which is the situation we have just described) then a signal may be sent from point B to point A so as to arrive at A before leaving B. A situation can be contrived whereby the returning signal from B arrives at A *before the original signal left A*. This raises a monstrous paradox. Not only would it be possible for an observer at A to have prior knowledge of an event which is to take place there in the future but—armed with this knowledge—he could take steps to prevent the event happening. Imagine that the archetypal Mad Scientist has constructed a device which will destroy the world, but which—when he presses the button, and *only* if he presses the button—releases a faster-than-light signal. He will abandon his Fiendish Plan to destroy the world only if he receives this signal prior to pressing the button. The paradox is clear: he will abandon the plan only if he receives the signal, but in order to receive the signal he must go ahead with his plan!

This is precisely the kind of paradox which would arise if tachyons exist *and can be used to convey information*. If tachyon transmitters and receivers can be used to pass messages between different observers then they can be used to signal to a given observer information about events in his future; or, putting it another way, it would be possible for someone to signal into his own past. Cause could precede effect and a host of logical contradictions would arise. These logical difficulties need not preclude the possibility that tachyons exist, provided that they cannot be used to convey information.

The same kind of paradoxes would arise if individuals were able physically to travel back in time and exert an influence on past events. A time traveller would then be able to travel back to the appropriate epoch in the past and take steps to prevent his own birth! If, in a fit of depression, he wished that he had never been born, then he would be able to have his wish; but then how could he go back to prevent his own birth . . .

In order to avoid logical contradictions such as these, we have to accept that faster-than-light signalling, the communication of information from the future to the past, and time travel into the past are all prohibited by the way in which the Universe is constructed. Travel into the distant future of the Earth is possible for relativistic astronauts because of time dilation, but the reverse operation is not possible. There is, then, a fundamental asymmetry in the nature of

Radio telescopes can "see" far further than optical ones, even those as large as the 200in reflector at Mount Palomar illustrated on page 189; they can thus look even further into the past of the Universe. (*Above right*) Karl Jansky (1905–1950), the US radio engineer who accidentally discovered that the centre of the Galaxy was a source of radio waves while attempting to eliminate static from a telephone system, shown here with his directional radio aerial system, the precursor of today's radio telescopes. (*Below right*) The modern version of Jansky's instrument, the 210ft (64m) radio telescope at Parkes, Australia, with a secondary telescope for interferometry work in the foreground.

time: we can remember the past, but we cannot receive information from the future; the past can influence the future, but the future cannot influence the past. We shall return to the problem of the "one-way" nature of time later in this chapter.

Spacetime

In everyday experience we are used to thinking of a world of three dimensions. Solid objects have length, breadth and height, each of these quantities being measured in a direction at right angles to the plane of the other two. As we have seen, we can describe position in space or the dimensions of a solid body by making reference to a Cartesian system of coordinates (Fig. 1, page 164) which measures distance in three mutually perpendicular directions (length, breadth, height, or x, y, z). We can visualize spatially extended objects in three dimensions, and we can visualize a particular location in three-dimensional space. For example, we can locate a particular office in a multi-storey office block by going "up", "along" and "across"; we take the elevator to the correct floor, walk along the corridor, then turn left or right at the appropriate door. However, most of us side with Newton and regard time as being something distinct from and unrelated to space, something which flows past at a steady rate.

Nevertheless, the idea of regarding time as a fourth dimension, in some way analogous to the dimensions of space, should not really strike us as being odd. After all, material objects have length, breadth and height, *and they exist for a finite period of time*. A freshly baked cake is a case in point. It has three finite spatial dimensions, and when it gets from the oven to the table its subsequent duration in time is strictly limited! Material objects—cakes, people, stars and planets alike—have finite extensions in space and finite extensions in time. It is quite reasonable, then, to think of time as being the fourth dimension.

It is easy for us to visualize time as a fourth dimension which has a character of its own and which is wholly independent of the three dimensions of space. What is not so immediately obvious is the interdependence of the four dimensions which emerges from the relativistic point of view. The Russian mathematician Hermann Minkowski first clearly spelled out this intimate relationship in 1908. In his own words, "Henceforth space by itself, and time by itself, are doomed to fade away into mere shadows, and only a kind of union of the two will preserve an independent reality."

In the pre-relativistic view of the Universe, spatial distances (e.g., length) and intervals of time were absolute quantities, and so were unaffected by the uniform relative motion of observers. As we have seen, according to special relativity, this is not so: spatial distances are affected by length contraction, and intervals of time are affected by time dilation. Different observers will *not* agree on lengths and times. Minkowski showed that it was possible to define an *interval* in four-dimensional spacetime which would be agreed upon by all inertial observers (i.e., all observers in uniform relative motion). Likewise, they would agree about the extension *in space-time* of a material object. The observers would disagree on the extension in time, and on the extension in space, but they would agree about some suitable combination of the two. The spacetime extension of an object, or the spacetime interval between two events, was an absolute quantity in Minkowski spacetime. Observers in uniform relative motion would see different projections of this spacetime interval in space and in time, depending upon their velocities; greater or lesser projections in space would be associated with lesser or greater projections in time.

We are familiar with this kind of idea so far as material objects are concerned, for objects take up different apparent shapes depending upon the angle from which they are viewed. For example, an oblong box may appear as a rectangle if viewed from the side, but as a square if viewed from the end. Looking a little more closely at this idea (Fig. 9a), let us think of the line which runs from the point A on the vertical (or "*y*") axis to the point B on

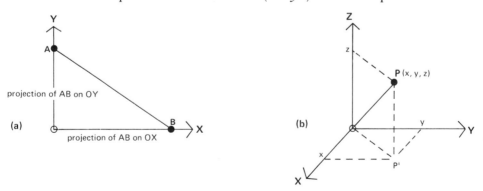

Fig. 9 Projections. In (*a*) we have a rectangular Cartesian frame of reference (see page 164) in two dimensions. OB is the projection of AB onto the axis OX, while OA is the projection of AB onto the axis OY. From Pythagoras' theorem we know that $AB^2 = OB^2 + OA^2$. In (*b*) we are thinking in terms of three dimensions. OP' is the projection of OP onto the horizontal plane, and OZ is the projection of OP onto the vertical axis. We can see that $OP^2 = OP'^2 + OZ^2$. Now, since $OP'^2 = OX^2 + XP'^2$, and $XP' = OY$, we see that $OP^2 = OX^2 + OY^2 + OZ^2$.

the horizontal ("x") axis. This line has length AB, but its projection on the x-axis is OB and its projection on the y-axis is OA. The theorem of Pythagoras—which haunted the schooldays of many of us—tells us that the square on the hypotenuse of a right-angled triangle is equal to the sum of the squares on the other two sides; i.e., $AB^2 = OB^2 + OA^2$ (where AB signifies the *length* of line AB, etc.). A distance in three dimensions can be described in a similar way. Thus the distance OP in Fig. 9b is given by $OP^2 = OX^2 + OY^2 + OZ^2$, or if we regard P as a point having coordinates x, y, z, then $OP^2 = x^2 + y^2 + z^2$.

Of course, we cannot *visualize* a four-dimensional situation, but Minkowski was able to show that, if the spacetime interval, s, between two events were described by $s^2 = c^2t^2 - (x^2 + y^2 + z^2)$, where c denotes the velocity of light, then all observers in uniform relative motion would arrive at the same value of s from their measurements of x, y, z, and t. The first part of the right-hand side of the expression represents the temporal "projection" of the interval, while the second part denotes its spatial "projection". If the spacetime interval is described in this way then its value is quite unaffected by the Lorentz transformation, even though the individual values of the temporal and spatial parts are altered by relative motion.

Minkowski called a point in spacetime a *world-point*, and the life history of a particle was represented by a line in spacetime called a *world-line*; extended bodies were represented by world-tubes. Minkowski was of the opinion that physical laws could be represented by the relations between world-lines of particles. According to this view, time was seen to be similar to space. The sum total of all possible world-points made up what Minkowski called "the world", and the four-dimensional world had an absolute nature, for *all* observers in uniform relative motion would agree on the spacetime intervals between world-points. Philosophically, this marked a fundamental new viewpoint. The idea of the flowing of time made no sense in four-dimensional spacetime: spacetime just "is"; it does not flow or change. All possible events exist in spacetime, and we as individuals happen to encounter these events; the passage of time, of which we are so acutely aware, seems simply to be a feature of our consciousness. Events themselves do not pass through time, time does not flow past events, and a universal "now" is a meaningless concept.

Einstein was quick to accept this four-dimensional view of the world about us, for special relativity clearly embodies the idea that

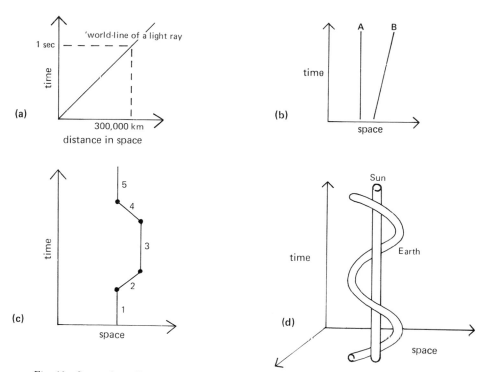

Fig. 10 Spacetime diagrams. In (*a*) time is measured in the vertical direction and space horizontally; the scale is such that 1 second has the same length as 300,000km, the distance travelled by light in 1 second. The path, or "world-line" of a ray of light is thus represented on the diagram by a straight line at an angle of 45° to both horizontal and vertical. (*b*) The world-line of a stationary particle is a vertical line (A), for its spatial position does not change. The world-line of a particle moving at constant velocity is a straight line (B) at an angle of less than 45° to the vertical (for the angle to be greater than 45° the particle would have to be travelling faster than light). In (*c*) we see a particle moving around: at stage (1) it is at rest; then (2) it moves off at a constant velocity until it again stops (3); after a time it returns (4) to its starting position (5). In the rather more complex example of (*d*) we represent space by the horizontal plane and time again vertically. Imagining the Sun, an extended body, to be stationary we may represent its "world-tube" by a vertical tube on the diagram. The Earth, which we regard as moving around the Sun as "time flows by", is denoted by a thinner tube which follows a helical path centred on the Sun's world-tube.

the properties of space and time cannot be considered in isolation from one another.

Although we cannot visualize in four dimensions, and we cannot possibly draw four dimensional spacetime on a flat, two-dimensional piece of paper, we can nevertheless make an adequate representation by means of a spacetime diagram (Fig. 10a), in which distance in space is measured horizontally and time is measured vertically. It is usually convenient to adjust the horizontal and vertical scales so that one second measured vertically has the same length as 300,000 kilometres measured horizontally. Since the speed of light is 300,000 kilometres per second this means that the world-line of a ray of light is represented by a straight line at an angle of 45° to the vertical (i.e., after 1 second of time the ray has travelled a distance in space of 300,000 kilometres). On such a diagram, the world-line of a stationary particle is represented by a vertical straight line (i.e., the position of the particle does not change with time), and the world-line of a particle moving at a constant velocity is denoted by a straight line inclined at an angle to the time axis (Fig. 10b). Since no material object may travel at the speed of light, the world-lines of material particles must always be tilted at an angle of less than 45° to the vertical on a diagram such as this.

Fig. 10c illustrates the world-line of a particle, initially at rest, which then moves to another position in space, remains there for a while and then returns to its initial location; perhaps it represents a day in the life of a commuter! The world-tubes of the Earth and Sun are depicted in Fig. 10d.

We can extract something rather striking from our spacetime diagram (Fig. 11) if we draw in the world-lines of light rays from world-point P into the future and if we also extend them back into the past. Since no material object or particle can achieve the speed of light then the world-line of any particle which was present at P must lie *within* the light-lines (or within the light-cone as we have drawn the diagram). Without exceeding the speed of light, an observer who was present at P could later be present at any subsequent event in the upper half of the light cone, and he could have been present at any event in the lower half of the cone. Since P may be connected to any point within the light cone by a signal which travels at a velocity less than or equal to the speed of light, then all observers in uniform relative motion will agree on the order of events relative to P which take place within the cone. The inside of the future light cone represents the *absolute future* of P (i.e., the sum of all possible events at which an observer present at P could subsequently be present), and the inside of the past light cone represents the *absolute past* of P (i.e., all those possible events at which an observer could have been present if he is now present at P).

The region outside the cone is that part of spacetime which cannot be reached by an observer who was present at P since he is not permitted to exceed the speed of light. Likewise, no causal influence can be exerted by P on any event in the outside region and, vice versa, no causal influence can be exerted upon P by any

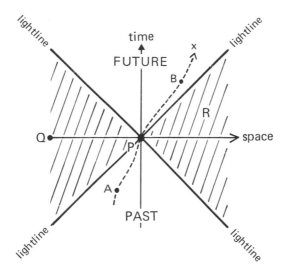

Fig. 11 The light cone and the absolute past and future. The world-lines of light rays through the point P are extended into the future and traced back into the past. A material particle, or observer, at P must lie in the future within the light lines (it cannot move as fast as light) and must in the past have been located between the light lines (for the same reason). The upper half of the cone formed by the light lines specifies the absolute future of particles present at P; the lower half their absolute past. The world-line of a particular particle is indicated by the line *x*. All observers in uniform relative motion will agree on the *order* of events A, P and B which take place along this world-line even although they may not agree about the intervals of time between them.

  The shaded region represents the region of spacetime inaccessible to any particle or observer present at P. Events Q and R cannot exert any influence either on P or on each other, for in order to do so they would have to be linked by a faster-than-light signal, which would be represented by a line inclined at more than 45° to the vertical.

event Q which occurs in that region. Furthermore, no causal influence can pass between Q and R, since that, too, would require a signal to travel faster than light. Because events outside the light cone cannot be causally connected to P then different observers in uniform relative motion will disagree on the order of these events; we have already looked at an example of the circumstances under which reversal of time order is possible (page 191).

The reversal of the order of events which exert no causal influence on each other causes us no problems in that we are not faced with the situation of effect preceding cause. However, if tachyons exist and if they can interact with normal matter in such a way as to communicate information or influence, then we would have problems. Then it *would* be possible for two events which have no definite time order as seen by observers in uniform relative motion to influence each other, being connected by a faster-than-light signal. As we have already seen, this leads to the paradox whereby it would be possible to know the outcome of an event before it occurred and to take steps to prevent its occurrence—a clear logical contradiction. As we have noted, we must assume for the moment that, even if tachyons do exist, they cannot be used to communicate information faster than light. If cause must always precede effect then there can be no "sub-space communication channel" between starships and "Starbase".

The light cone, then, divides all of spacetime into two distinct regions, the region inside the cone which is accessible to P, and the region outside, which is inaccessible to P. Events which happen inside the cone can be unequivocally dated as happening "before" P or "after" P, while events outside the cone have no definite time order. World-lines from P to points inside the cone are called *timelike* because the temporal part of the interval* between two events on the world-line is greater than the spatial part, the paths of light rays are called *null* (since the spatial and temporal parts are equal, the interval between two points on a light line is zero), and hypothetical trajectories outside the cone are called *spacelike*, because the spatial part of the interval is greater than the temporal part. Material particles may pursue only timelike trajectories.

The General Theory of Relativity

As we have seen, special relativity embodies the idea that the properties of space and time cannot be considered in isolation from one another. But special relativity is restricted in its applicability to inertial observers, i.e., to observers in uniform relative motion. For a body to move at a uniform velocity it has to be free from the influence of all forces, for forces give rise to accelerations. In particular, all bodies are influenced by gravity, and there is nowhere in the Universe where the effects of gravitation are completely absent. By moving away from the Earth, the gravitational attraction of the Earth may be reduced to as low a value as we wish, but can never be reduced absolutely to zero. In any case, as we move away from the Earth, we are still subject to the gravitational influence of the Sun. By moving far enough away from the Sun, we can make its attraction negligible, but we find we are still subject to the general

* Interval: earlier we defined the spacetime interval, s, by
$$s^2 = c^2t^2 - (x^2 + y^2 + z^2),$$
the first part of the right-hand side being the temporal part and the second the spatial part. If the distance $\sqrt{(x^2 + y^2 + z^2)}$ is less than the distance ct which can be covered by a ray of light in time t then s is greater than zero and the world-line is timelike.

gravitational influence of the Galaxy, and so on. Special relativity, then, is really only valid in the absence of gravity, and is an approximate theory where gravity is present. Now, in fact, in most circumstances gravity is a rather weak force, and special relativity works exceedingly well; it certainly is more satisfactory than Newtonian mechanics, and that itself is perfectly good enough for most practical purposes. Nevertheless, Einstein wished to extend relativity to include *accelerated* observers—i.e., observers who were subject to forces—and so it was that he developed the *general theory of relativity* which, published in 1915, turned out to be a superior theory of gravity to that of Newton. For 99.99% of applications Newton's theory is perfectly adequate, but there are situations in which the old-established law of gravity is inadequate, and it is then that general relativity comes into its own.

General relativity takes the concept of mutable time a stage further, as we shall see, but first let us look at the foundations of this monumental theory. The fundamental new principle upon which this new theory was to rest was the *principle of equivalence*,

GRAVITY   ACCELERATION

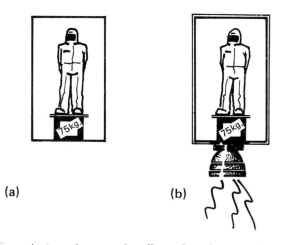

(a)   (b)

Fig. 12 The equivalence between the effects of gravitation and acceleration. The principle of equivalence asserts that an observer inside a sealed box cannot distinguish between the effects of gravitation and those due to acceleration of the sealed box. For example, a man in the box on the Earth's surface stands on a spring balance which registers his weight as 75kg (a). He could equally well be out in space in a box which is being accelerated uniformly at a rate equal to the acceleration due to gravity at the Earth's surface (9.81m per second per second). In case (b) also the spring balance would register the fact that he was pressing down on it with a force of 75kg, and he would feel the sensation of weight just as if he were on the Earth's surface.

enunciated by Einstein in 1907, eight years before the full general theory was published. The principle asserts that there is no way of distinguishing locally between the effects of gravity and those effects generated by acceleration. An observer in a sealed box (Fig. 12) should not be able to tell whether his experience of "weight" arises because the box is sitting on the surface of the Earth where he is subject to the Earth's gravitational attraction, or because he is far out in space, well away from any attracting body, in a box which is being accelerated (perhaps by a rocket motor attached to it). The force experienced will feel the same in either case, and there is no measurement which he can make *inside* the box which will allow him to tell the difference.

This principle is an extension of the observed proportionality between gravitational and inertial mass. Earlier we mentioned Newton's laws of motion. According to Newton's second law, the acceleration a produced in a body of mass m by a force F is given by $a = F/m$; acceleration equals force applied divided by the mass of the body being accelerated. "Mass" in this context is a measure of the body's resistance to acceleration; i.e. to its *inertia*, and is therefore known as "inertial mass".

According to Newtonian gravitation, a mass m and a mass M will attract each other with a force equal to GmM/d^2 where G is the gravitational constant, and d is the distance between the bodies. Mass in this case is a measure of "the quantity of gravity" possessed by that body; it is analogous to the quantity of electrical charge possessed by an electrically charged body. Now the curious thing is that, whereas a body can, within reason, possess any value of electrical charge, so that its acceleration in an electrical field can have different values depending upon the value of charge, all material objects—whatever their gravitational mass—are accelerated by precisely the same amount in a gravitational field. Heavy bodies fall with precisely the same acceleration as light bodies. This is a result which was shown experimentally by Galileo (there is a story, probably apochryphal, that he demonstrated this in public by dropping different weights from the Leaning Tower of Pisa, showing that the different weights took the same time to reach the ground), and it may readily be demonstrated by the reader dropping heavy stones and light pebbles and noting the outcome.

On the face of it, this may seem a surprising result and you may argue that it doesn't work with a balloon or feather and a pebble, for the pebble will fall much faster than either of the other two objects. However, this is not a valid objection, since it is the resistance of the Earth's atmosphere which slows down the motion of the balloon or feather. A dramatic public demonstration of this took place on the surface of the Moon during the Apollo 15 mission. An astronaut stood before the television camera and dropped a feather and a geological hammer: as expected, they both hit the surface of the Moon at the same time!

It would appear, then, that gravitational mass and inertial mass of a body are always in precisely the same proportion, and that the inertial mass (which determines acceleration) and gravitational mass (which determines gravitational attraction) are wholly equivalent. A more massive body experiences a stronger gravitional attraction exerted on it by the Earth than does a less massive body, but—because its inertial mass is greater in exactly the same proportion—the acceleration which it experiences is exactly the same as that of the less massive body. In effect, the extra force due to the extra mass is cancelled out by the extra inertia of the body. Units of measurement of inertial and gravitational mass are usually chosen to make the two equal, and this equality has been tested to a very high degree of precision in experiments carried out by, for example, Newton, Eötvös, and, more recently, by Dicke and Braginskii to an accuracy of one part in a million million.

Returning again to the notion of observers in closed boxes, let us imagine an observer standing inside a lift, holding a pile of objects in his hand (Fig. 13). If the lift cable should snap, the lift will fall freely under the influence of gravity, accelerating at a steady rate, and so will its contents: the observer and all the other objects in the lift will fall at precisely the same rate as the lift itself. The falling person will feel no force between his feet and the floor of the lift, and so will experience the state of weightlessness; all the objects he was holding in his hand will appear to float in mid-air. Of course, this dream-like state of affairs will be brought to an abrupt and permanent end when the lift hits the bottom of the shaft, but that does not invalidate the point that, while the observer and his container are in free fall, he experiences no effects attributable to a gravitational field. The same experience occurs for an astronaut in a spacecraft in orbit round the Earth; such a craft is moving freely under the influence of gravity, and its inhabitants feel no weight.

We have a two-way equivalence between gravitational effects and acceleration effects. A force indistinguishable from gravity is experienced by an observer who is being accelerated (say, an astronaut when the rocket motor of his craft is firing) but this force

(a) **(b)**

Fig. 13 Free fall and weightlessness. In (*a*) someone is standing in a stationary lift holding a tray laden with objects. He feels his normal weight, and so do they. But in (*b*) disaster strikes. The cable has snapped, allowing the lift to fall freely under the influence of gravity. The lift and its contents fall at precisely the same rate, accelerating in unison. Neither the unfortunate "someone" nor the objects which he carried experienced any sensation of weight, and so they are free to float around within the contents of the lift. Until . . .

vanishes when the accelerating force is removed (i.e., when the motor is switched off). Likewise the effects of gravity may be eliminated, "switched off" or "transformed away", by allowing the observer's box to fall freely under gravity. Apparent "gravitational" forces are familiar to us these days, and it has been suggested that artificial gravity may be generated in a large space station by the simple expedient of setting it spinning. People on the inside rim of the station (assuming a symmetrical space station) would be subject to a constant acceleration since they would be constrained to move in a circular path by the rotation of the station instead of moving through space in a straight line at uniform velocity. The direction of the acceleration is towards the centre of the circle, but, because of their inertia, they will feel themselves pressed against the outside wall by a "force" indistinguishable from gravity: the apparent force which they experience is commonly referred to as centrifugal force. The strength of artificial gravity may be adjusted by varying the rate of rotation, or an individual may select the level of gravity he desires simply by moving closer to or further from the axis of rotation: the closer he gets, the weaker the apparent force of gravity becomes, and at the axis it disappears altogether.

It is important to note, however, that the equivalence principle applies only within closed volumes of space, where gravity may be regarded as constant. In the case of a large box resting on the surface of the Earth, tidal effects may become apparent: the top of the box will be significantly further from the centre of the Earth than is the bottom of the box, so that objects released near the top of the box will accelerate less rapidly than objects released near the bottom of the box. (This is true also in a small box, of course, but the effect would be too minor to notice.) Likewise, objects released at different points inside a large box in orbit around the Earth will fall freely at different rates, so that an object in the lower part of the box will slowly drift further away from an object in the upper part. A similar effect would arise from the fact that each particle inside a freely falling box would accelerate along a line directly towards the centre of the Earth. Thus particles floating on opposite sides of a freely falling lift would drift towards each other as the lift approached the centre of the Earth.

Einstein argued that one could always choose a region small enough (i.e., a small enough box) in a state of free fall in a gravitational field, within which the effects of gravitation could be eliminated, or within which the gravitational field could be regarded as being uniform. In this way we get back to the premise that the local effects of gravitation are indistinguishable from those of acceleration. He proposed that a freely falling box had the same status, locally, as an inertial frame, and so he regarded such a box as a "local inertial frame". All the laws of physics as determined by experiments carried out inside such a box should be the same as those determined by observers in uniform motion. Expressed in these terms, his equivalence principle, so stated, extended the relativity principle beyond the restricted application—which it has in special relativity, to observers in uniform relative motion, and showed it to be applicable to accelerated frames of reference, too. Provided that one restricted one's attention to a region of space-time sufficiently small that the effects of gravitation were effectively constant, then the laws of physics should be the same for all observers *whatever* their state of motion.

We can readily see some of the consequences of the equivalence principle by returning once again to the analogy of the inhabitants of a freely falling lift (Fig. 14a). By the principle of equivalence, no effects of gravity should be apparent to the occupants, so that if A decides to throw a ball to B the ball will move in a straight line from A to B. If the lift had walls made of one-way glass, so that

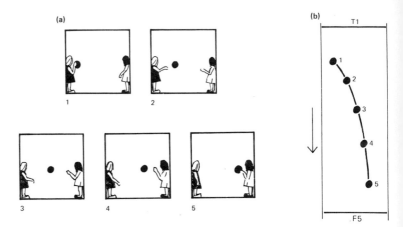

Fig. 14   The path of a particle in a gravitational field. In (*a*) observers A and B
are inside a freely falling lift; A throws a ball to B and, because the effects of
gravity are not apparent inside the lift, the ball follows a straight-line trajectory.
Stages (1) to (5) are separated by equal intervals of time. From the point of view
of an outside observer, stationary with respect to the lift shaft, however, the points
(1) to (5) in (*b*) denote the observed positions at the equivalent stages in (*a*). The
discrepancy in the observations arises because the lift and its contents are
accelerating downward relative to the outside observer. T_1 represents the top of
the lift at stage (1) and F_5 the bottom of the lift at stage (5).

outsiders could see in, but insiders could not see out, then the outsiders' view would be as represented in Fig. 14b. Relative to the
background of the lift shaft, the lift and its contents would be accelerating downwards and, if the position of the ball in its flight from
A to B were plotted, it would be seen to be following a parabolic
trajectory, just as we would normally expect a projectile to do if it
were thrown by someone standing on the Earth's surface.

Now let us suppose that A decided to send a ray of light (from a
laser perhaps) across to B. Within the confines of the lift, the time
of flight of the light ray would be very short indeed, but finite none-
theless. A and B would agree that, as expected, this beam of light
followed a straight line path between them; but an outside observer
would reach a different conclusion. Because of the acceleration of
the lift during the brief travel time, that ray would have followed a
curved path in the outsider's frame of reference. The curvature
would be very slight since the velocity of light is very great, but
there is no doubt that the path of the light ray should be curved;
the conclusion to be drawn from the equivalence principle is that
rays of light should be curved in the presence of a gravitational
field. In fact, the equivalence of mass and energy which stems from
the special theory of relativity suggests that light (which carries

energy), like mass, should be affected by a gravitational field, so this result should not be too surprising.

The bending of light in a gravitational field was one of the crucial predictions of Einstein's general theory, and was confirmed by astronomical measurements made by Sir Arthur Eddington in 1919. A ray of light from a distant star passing close to the edge of the Sun should be deflected by a small but measurable amount (Fig. 15). The problem about making such measurements is, of course, that stars cannot be seen in daylight. During a total eclipse, when the Moon blocks out the brilliant disk of the Sun, then stars do become visible and it was during a total eclipse that Eddington measured the apparent change in position of stars close to the Sun in the sky and showed that the bending of light agreed with the predictions of the theory. More recently, it has been possible to measure the deflection of radio waves passing the edge of the Sun, and this too accords with the theory.

The other key factor in the setting up of general relativity was Einstein's linking of gravitation to the *geometry* of spacetime. Going back to the spacetime diagram (Fig. 10), we see that the world-line of an inertial observer is a straight line. However, since we cannot escape from gravity in the Universe, all particles must be subject to some degree of acceleration (however small) and the world-line of an accelerated particle is curved (Fig. 16). By restricting our atten-

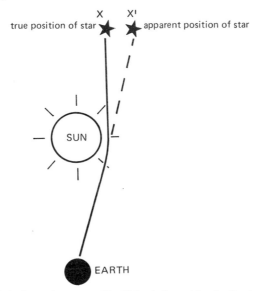

Fig. 15 A ray of light from the star at X will be deflected by the Sun's gravitational field if it passes close to the edge of the Sun. An observer on Earth will thus see the star shifted away from its true position to the apparent positon X'.

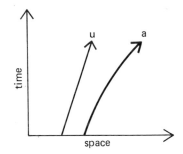

Fig. 16 A spacetime diagram representing acceleration. The world-line of a particle (or observer) moving at uniform velocity is indicated by the straight line *u*. The world-line of an accelerated particle, however, is curved, *a*, sloping further away from the time-axis as the velocity increases, although never, of course, becoming inclined at an angle greater than 45° to the vertical.

tion to sufficiently small volumes of spacetime we can produce local inertial frames within which uniform straight-line motion is possible within the limits of the accuracy of measurement, but when we look to larger and larger volumes of spacetime we become aware of a progressive distortion of these world-lines. This is similar to the situation which we encounter on the surface of the Earth: within a sufficiently small area we can regard the surface of the Earth as being flat—a surveyor setting up a building site for a house has no need to take into account the curvature of the Earth, but an airline pilot or round-the-world sailor is acutely aware that the Earth is curved!

The crucial step taken by Einstein was to suggest that gravitation, which was responsible for world-lines being curved, was a property of spacetime itself. In other words, spacetime is curved in the presence of massive bodies. Spacetime in the vicinity of a massive body would be strongly curved, and it is this curvature of spacetime which we interpret as a gravitational field. Far from any matter spacetime would be very nearly flat, and the geometry of spacetime would be, near enough, the plane geometry of Euclid, but in any other situation we have to use the geometry of curved spaces and surfaces developed in 1854 by the German mathematician G.F.B. Riemann.

In ordinary plane geometry we are used to the idea that the shortest distance between two points is a straight line. In Riemannian geometry the shortest distance between two points is a curved path known as a *geodesic*. For example, one cannot draw a straight line on the surface of a sphere, such as the Earth, and the shortest distance between two points on the surface of a sphere is a curve which is part of a great circle (a great circle is one whose centre is

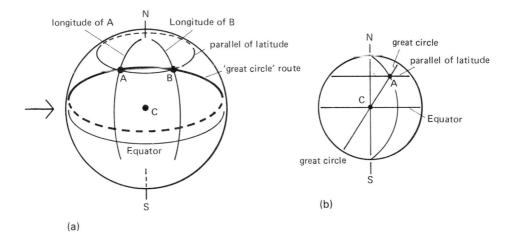

longitude of A
N
Longitude of B
parallel of latitude
'great circle' route
A
B
C
Equator
S

(a)

N
great circle
parallel of latitude
A
C
Equator
great circle
S

(b)

Fig. 17 The shortest distance between two points on a sphere. In (*a*) A and B are two points on the Earth's surface which lie at the same latitude. The shortest distance between them is measured along the arc of the *great circle* passing through A and B: a great circle is one whose plane passes through the centre of the sphere on which it is drawn (in this case the centre of the Earth, C). The shortest route is *not* the route which follows the "parallel" of latitude between A and B. In (*b*) we see a cross-section of the Earth viewed from the direction of the arrow in (*a*).

at the centre of the sphere; e.g., the equator or a circle of longitude is a great circle, but a "parallel" of latitude is not). Thus the shortest route between two points widely separated but lying at the same latitude is not to travel along that line of latitude but to follow a curve as shown in Fig. 17. To travel between two points at latitude 45° but located on opposite sides of the Earth along a parallel of latitude involves covering a distance of about 10,000 kilometres, but by following a great circle route which passes directly over the pole only some 6,700 kilometres need be covered!

In Einstein's view, freely falling particles (including rays of light) followed world-lines which are geodesic in curved spacetime. Rays of light follow the "straightest possible" paths, but their world-lines too are curved, being known as *null-geodesics*. Thus a planet in orbit around the Sun is in free fall round the Sun and is following a geodesic in spacetime. A planet is not held in its orbit by a force (Newtonian gravitation) exerted by the Sun, but because it is following its natural path in the curved spacetime which surrounds the Sun. Gravity, then, is seen as the curvature of space-time, not as a force which (somehow) acts instantaneously across space in the way Newton had envisaged.

One prediction of the general theory of relativity has already

209

been mentioned, namely the deflection of light rays in a gravitational field, and this has been tested experimentally to a fair degree of accuracy. A second test was provided by the motion of the planet Mercury, the nearest planet to the Sun, which moves on an orbit which is markedly elliptical. The point of closest approach to the Sun is, as we saw in Chapter 3, known as *perihelion* . It was already known from astronomical observations that the perihelion of Mercury slowly rotates around the Sun; i.e., the orbit itself slowly shifts in space. Much of this motion could be explained by Newton's theory of gravitation as being due to the disturbing effects of the other planets, but there remained a discrepancy of 43 seconds of angular measurement *per century* which could not be accounted for. Einstein's theory was able precisely to explain the motion of the planet.

What is the point of all this discussion about the nature of gravity in a book which is devoted to the subject of time? Since gravitation is interpreted as the curvature of spacetime, then we should expect temporal effects to occur in the vicinity of massive bodies and, indeed, there is a further prediction of general relativity which has direct relevance to our ideas of time: *natural clocks run more slowly in a strong gravitational field than they do in a weak field*. In other words, there should be a gravitational time dilation analogous to the time dilation which results from relative velocity in special relativity. A clock placed on the surface of a massive body will run more slowly than an identical clock placed far away from that body. In the case of the Earth, the effect will be very small; the rate of a clock far from the Earth, according to the theory, will differ from that of an identical clock on the Earth's surface by only seven parts in ten thousand million; i.e., the Earth-based clock will run slow compared to the other by only some 20 seconds in a millennium. Where very strong gravitational fields are involved, however, the effect may be dramatic indeed.

Time dilation is directly related to the observable phenomenon of the gravitational redshift. Light coming out of a strong gravitational field has to work hard to climb away from the surface of a massive body, and, in so doing, loses energy. The lower the energy of a photon (a "particle" of light), the longer the observed wavelength of that light; the loss of energy corresponds to an increase in wavelength, and the wavelength reaching a distant observer is longer than the wavelength emitted at the surface of the body. In other words, the light reaching a distant observer is redshifted. The gravitational redshift refers to a change in a natural clock, the natural clock being the interval between successive crests of a light wave. Redshifted light has a longer wavelength so that fewer wavecrests per second reach the observer (i.e., the *frequency* of the light is reduced), and the time interval between successive crests, therefore, is increased. The successive wavecrests could be regarded as the successive ticks of a clock, so that a clock in a strong gravitational field would be seen by a distant observer to be running slow compared to his own local (or "proper") clock.

The gravitational redshift has been observed in the light coming from white dwarf stars. These are stars in the final stages of their lives, having shrunk to a fraction of their former dimensions. A typical white dwarf has a mass the same as that of the Sun but is only about the size of the Earth; consequently the material of which it is made is so dense that a teaspoonful—if it could be brought to the Earth—would weigh several tonnes. The force of gravity at the surface of such a star is several hundred thousand times greater than the force of gravity here on Earth, and in these circumstances the gravitational redshift is much more apparent.

Gravitational redshift and time dilation are direct consequences of the equivalence principle, so that measurements of these effects amount to tests of the principle. Until recently, the most precise measurement of the effect was that carried out by R.V. Pound, G.A. Rebka and J.L. Snider at Harvard in the 1960s. They measured the change in frequency of gamma rays between the top to the bottom of a 75-foot (about 23m) tower, and confirmed the predictions of the theory to an accuracy of about 1%. In 1976 a much more precise experiment was carried out by R.F.C. Vessot and M.W. Levine of the Harvard-Smithsonian Center for Astrophysics. In this experiment, a hydrogen maser clock was fired in a rocket to an altitude of about 10,000 kilometres, and the signals relayed from this clock were compared with identical clocks on the ground. While in flight, the rocket-borne clock was in a weaker gravitational field than the Earth-based clocks, and at its maximum altitude the frequency of that clock should have differed from that of the ground-based clocks by about 4.5 parts in ten thousand million. Although this is a microscopic difference, the clocks used for the measurements were considered to be accurate to about one part in one thousand million million (i.e., 1 in 10^{15}), and the predicted effect should have been well within the capabilities of the experiment.

Analysis of the results proved to be a difficult operation, but by 1978 the experimenters were confident that their results agreed with Einstein's predictions to an accuracy of two parts in 10,000; i.e. to within one fiftieth of 1%. Further analysis may improve the results even further. There seems no doubt, then, that the gravitational time dilation effect exists, and that the principle of equivalence holds good. This principle is central to the general theory of relativity and, as things stand at present, general relativity remains the best theory of gravitation which we have.

In summary, general relativity is a theory which welds together the equivalence principle and the concept that gravitation may be regarded as a distortion of the geometry of spacetime such that the world-lines of freely falling particles are geodesics (the closest analogy in curved spacetime to the idea of a straight line in plane geometry). The precise description of gravity in these terms is contained in the *field equations* which Einstein published in 1915. This elegant theory replaces the Newtonian idea of gravitational forces with the notion that bodies pursue their natural paths in curved spacetime; bodies do not "feel" gravitational forces, but respond to the curvature of spacetime in their vicinity. Where the curvature of spacetime is slight, the path of a particle is practically straight, but in strongly curved spacetime—near a massive body—its path is appreciably bent. It is implicit in this theory that accelerated motion ("free fall") is the natural state of motion in the Universe. Newton and Galileo had already overthrown the long-held idea that force was necessary to maintain motion and had replaced it with the view that the natural state of motion was uniform motion in a straight line, force being required only to *change* that state of motion. Einstein went further and showed that the natural state was to be in free fall, force being required only to change *that* state. The only force experienced by the occupants of a freely falling lift will be the impact which halts their fall at the bottom of the shaft.

Space, time and gravity are intimately entwined. Not only does velocity affect the rate of time's passage, but the strength of the gravitational field, or the corresponding distortion of spacetime, affects the rate as well. In special relativity we saw that the rate of a clock carried in, say, a fast-moving spacecraft is slower than the rate of a stationary clock in its vicinity. In general relativity we see also that the rate of a clock placed in a strong gravitational field is

(*Right*) The Crab Nebula, the remnants of a supernova recorded in 1054 by Chinese astronomers, a rapidly expanding shell of incandescent gas (*above*) at the centre of which lies a pulsar, a rapidly rotating neutron star (*below*). Matter in a neutron star is so dense that, according to one source, a thimbleful of it would, if brought to Earth, weigh the same as a fleet of ocean liners.

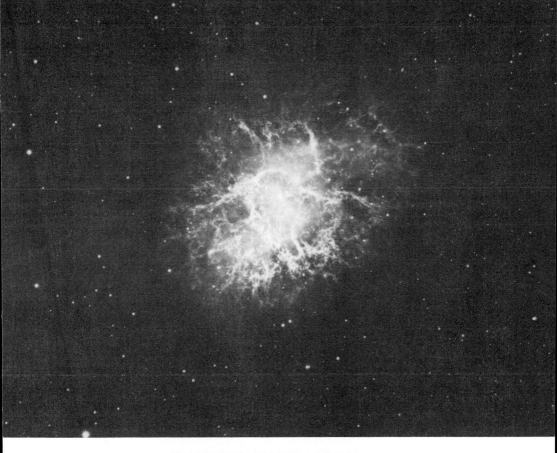

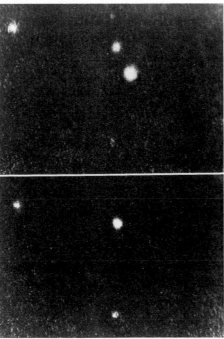

slower than the rate of a clock in the immediate vicinity of an observer located far from that gravitational field. The clock carried with an observer—i.e., the clock which keeps pace with him and remains by him whether he is moving at uniform velocity or is subject to acceleration—is his *proper clock* and it measures his *proper time*. Just as an observer aboard a fast-moving spacecraft will not be aware of any change in his onboard proper clock, so an observer placed in a strong gravitational field will not be aware of his proper clock running slow. In neither case will any change be discernable (at least by means of local observations). Observers in uniform relative motion, or in different gravitational fields, will disagree about the rate at which time is passing, but no one observer, whatever his state of motion, can claim to have a truer measure of time than any other. Once again, there is no absolute time; time itself is a variable, flexible commodity.

Black Holes and Time

As we have seen, in normal circumstances the effects of spacetime curvature are very slight and the most precise measuring techniques (such as those used in the experiment of Vessot and Levine) are essential to detect them. The lifespan of one member of a pair of twins who lived on the ground floor of a block of flats would not be measurably greater than that of the other member who lived on the top floor (not, at least, for reasons connected with general relativity). However, in very strong gravitational fields, we may expect these effects to become readily apparent.

There is great excitement in physics and astronomy today over the possibility of detecting entities known as *black holes* which, if they exist, represent regions of space in which the gravitational field (or spacetime curvature) due to collapsed matter is so great that not even light can escape. In 1916 the German astronomer Karl Schwarzschild published his solution of Einstein's field equations for the spacetime in the vicinity of a spherical lump of matter: the solution showed that if a mass M were compressed within a sufficiently small radius (now known as the Schwarzschild radius) R_s, then the distortion of spacetime would be so great that no signal of any kind could escape from within that radius. The value of the Schwarzschild radius is given by the simple formula, $R_s = 2GM/c^2$, where G is the familiar gravitational constant and c denotes the velocity of light. In fact, a similar idea had been discussed back in 1798 by the French mathematician Pierre de Laplace. Treating light as a stream of particles subject to the effects of gravity he suggested that there might exist in the Universe objects which were so massive that the escape velocities* at their surfaces were greater than or equal to the speed of light. That being so, he argued, light could not escape from such bodies and they would remain invisible. The value of the critical radius which he obtained was just the same as that obtained by Schwarzschild using general relativity. (In this case, Newtonian theory and general relativity gave the same result, but it should be pointed out that Laplace's reasoning did not lead to an entity with quite the same properties as a black hole in general relativity.)

Most objects in the Universe are much larger than their Schwarzschild radii—e.g., the Schwarzschild radius of the Sun is about 3 kilometres, and the actual radius of the Sun is about 700,000 kilometres; the actual radius of the Earth is some 6,400 kilometres and its Schwarzschild radius is in the region of one *centimetre*. There appears to be no way in nature by means of which either the Sun or the Earth could be compressed sufficiently to form a black hole. If black holes exist in the Universe it is most likely that they will have originated from the collapse of some of the most massive of the stars at the end of their life cycles.

A star such as the Sun is in a state of balance. The gravitational attraction of each particle on every other one acts inwards, tending to compress the star, but this force is counterbalanced by the pressure inside the star, sustained by the nuclear reactions which produce the star's energy. For as long as it can continue to generate energy, the star remains inflated like a balloon, but ultimately it must run out of fuel and, when it does so, it can no longer support its own weight. Most stars, the Sun included, are likely to end up as white dwarfs, stars which have been compressed by gravity to densities about a million times greater than that of water. Gravity cannot compress such a star any further provided that its mass is less than about 1.2 times that of the Sun.

If the final mass of the star does exceed this limit, then it may end up as a much more compressed object, known as a *neutron star*.

* A body projected vertically from the surface of the Earth will fall back to the Earth again unless its velocity exceeds a particular value known as the escape velocity; if it exceeds this value, the body will continue to move away and will never return. The value of the escape velocity V_e at a distance R from the centre of a body of mass M is given by $V_e = \sqrt{2GM/R}$, and at the surface of the Earth it has a value of about 11 kilometres per second. The critical radius within which a mass M must be compressed in order that (according to Laplace) the escape velocity at its surface be equal to or greater than the speed of light is obtained by setting $V_e = c$. Thus, $R = 2GM/c^2$.

Inside such a star, the whole structure of atoms has been destroyed; negatively charged electrons have been forced to combine with positively charged protons to form electrically neutral neutrons. A neutron star would be less than ten kilometres in radius, yet might contain more material than the Sun, compressed to a density so great that a teaspoonful of neutron star material, if brought to the Earth, would weigh a thousand million tonnes! Astronomers are now confident that they have detected neutron stars in space.

Theory suggests that any star whose final mass exceeds about two solar masses must collapse even beyond the neutron star stage. If such a star, or stellar remnant, begins to collapse under its own gravitational attraction then there is no force known to physics which can halt the collapse. In principle, all the matter of the star would be compressed into a point of infinite density, a point where gravitational forces would be infinitely great, a point known as a spacetime *singularity*. Before the star had collapsed to this extent, it would have passed inside its Schwarzschild radius and given rise to a black hole. Whatever happens to the material of the star thereafter, no information about its fate can reach the outside Universe for, to do so, a signal would have to exceed the velocity of light. The boundary of a black hole is called the *event horizon* for the very good reason that no information about any events which occur within the boundary can ever be communicated to the outside world. A ray of light emitted at the event horizon would remain there forever, neither moving out nor falling in, like Alice and the Red Queen, running flat out in order to stay at the same place; a ray of light emitted inside the event horizon would be dragged inexorably towards the centre of the black hole.

Curious things would happen to time in the vicinity of a black hole. Let us imagine an experiment whereby an intrepid astronaut agreed to drop into a black hole in the interests of science (volunteers might be hard to come by). He is equipped with a powerful laser torch which emits a pulse of light once every second, so that we can follow his progress from a safe distance by observing the flashes from his torch. The astronaut begins to fall towards the black hole, recording the time which elapses from the beginning of his mission on a clock which he carries with him, and emitting a pulse from his laser at one-second intervals as measured on this, his proper clock. From his point of view, he will accelerate rapidly towards the black hole, crossing the event horizon without noting anything unusual (we shall ignore here the tidal stresses to which

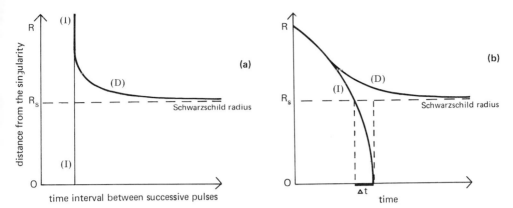

Fig. 18 Contrasting views regarding falling into a black hole. In (*a*) an intrepid astronaut (I) falls in towards a black hole; he is signalling back to a distant observer (D) at regular intervals according to his proper clock. According to D, as the astronaut gets closer and closer to the event horizon at R_s the time interval between the successive pulses he receives becomes longer and longer. In (*b*) we see the situation according to the astronaut. He continues to accelerate as he approaches the black hole and, after crossing the event horizon, he falls to the central singularity in a very short period of time, Δt (for a typical stellarmass black hole Δt might be about one ten-thousandth of a second). But according to D the astronaut never gets there, it taking an infinite time for him to cross the event horizon.

he would be subjected). For a black hole of stellar mass only something like one ten-thousandth of a second will elapse, as measured by his clock, between his crossing the event horizon and falling to the central singularity where he will be crushed out of existence by the infinite gravitational forces. This will be a very real and painful event for him and, to add insult to injury, he will not be able to tell anyone about his experiences once he has entered the black hole as his signals will not be able to escape.

We who remain at a distance from the black hole will see a quite different view of things. True, we will see the astronaut accelerate towards the black hole, but as he gets close to the event horizon (Fig. 18) something most peculiar will begin to happen. The time interval between successive flashes from his laser will become progressively longer and longer the closer he gets to the black hole; we should conclude that his clock is slowing down relative to our own proper clock, and this is just what we would expect for a clock placed in a powerful gravitational field. The time interval between successive pulses would become infinitely long as the astronaut reached the event horizon, and we would conclude (since in principle we could receive a pulse an infinitely long time after the start of the mission) that he never crossed the event horizon at all! From a distant observer's point of view, time would stand still at the event horizon of a black hole, but from the point of view of the in-falling astronaut, he would cross the horizon and fall to the centre in a very short time indeed.

Who is correct? As we have seen before, each observer is correct within his own frame of reference, but neither can claim that the other is wrong. The distant observer is correct in what he sees, but the in-falling observer is made painfully aware that his viewpoint is correct so far as his own frame of reference is concerned!

One might well argue from this that black holes could not form at all. After all, if we were to watch a star in the act of collapsing on itself, its timescale should slow down as it approached its own Schwarzschild radius and it would take an infinite time to cross its own event horizon. We should be able to see forever a "frozen image" of the star made up of light emitted as it reached the event horizon. However, gravitational time dilation and redshift go hand in hand: light emitted in close proximity to the event horizon should be so severely redshifted that its wavelength becomes infinitely long. In a tiny fraction of a second, as the collapsing star reached its Schwarzschild radius, its light would be redshifted out of sight and the star would become quite undetectable at any wavelength. Gravitational time dilation notwithstanding, the redshift ensures that collapsing stars and in-falling astronauts do indeed vanish—permanently.

Black holes, by definition, would seem to be undetectable. They absorb anything—radiation or material objects—which falls upon them, and emit nothing, for no signal or material object can pass outwards through the event horizon. However, a black hole still exerts a gravitational influence on its surroundings and it is this which may betray its presence. For example, there exists a source of X-rays in the sky which was discovered in 1970 by the US satellite Uhuru, and which is known as Cygnus X-1. It consists of a hot, massive star with an invisible companion which seems to be far too massive to be either a white dwarf or a neutron star; the mass of the invisible companion, deduced from the orbital motion of the visible star, seems to lie between six and fifteen times that of the Sun. Material appears to be flowing from the visible star towards the invisible object. What would be expected to happen in a situation where a conventional star and a black hole were in close proximity to each other, orbiting under their mutual gravitational attraction, is that matter flowing from the visible star would form a circulating disk of gas outside the event horizon of the black hole. The gas falling in towards the event horizon would be so intensely heated that it would emit X-rays, and this may explain the X-rays observed to come from Cygnus X-1. The X-rays in this case are assumed to be coming from material *outside* the event horizon; no signal can come from within that boundary.

The evidence to suggest the presence of a black hole in Cygnus X-1 is good, but it is not conclusive. Nevertheless, a very similar X-ray binary, V861 Sco, has been discovered in the constellation of Scorpius, and the May 1978 issue of *Astrophysical Journal* carried two papers describing observations which were consistent with the idea that the giant elliptical galaxy, M87, contained a truly massive black hole containing about as much material as five thousand million suns! There is no reason, in principle, why super-massive black holes should not exist, and many astronomers have suggested that such entities may lurk at the centres of galaxies (even, perhaps, at the centre of our own Galaxy) acting as a "powerhouse". A massive black hole onto which material was falling could act as a powerful energy source, the energy being released by in-falling material *before* crossing the event horizon. Compact energy sources like these would provide a most convenient explanation of peculiar objects such as quasars—compact, powerful, and immensely distant objects lying well beyond the confines of our own Galaxy. There is no definitive proof that black holes have been detected as yet, but the circumstantial evidence is certainly mounting up in a convincing way.

An interesting aspect of the mathematics which describes space-time in the vicinity of a black hole is that it seems as if a black hole connects the spacetime of our Universe with a wholly different spacetime. It is as if there is "another universe" on the "other side" of a black hole. As the black hole is approached, the curvature of spacetime increases up to the event horizon, beyond which we cannot see; however, if we continue to follow the curvature of spacetime it appears to open out again into another "flat" space-time. Does this really imply that a black hole connects our Universe to another one of which we have no awareness in any other way, or—as some have suggested—is spacetime as a whole distorted in such a way that a black hole links two widely separated regions of spacetime in our own Universe. A hypothetical connection between two regions of spacetime is called an *Einstein-Rosen bridge* or *wormhole*. Could matter which disappears into a black hole reappear in another universe, or even in a different part of our Universe? Does matter which falls into a black hole emerge "else-where"? All these possibilities have been discussed.

However, the investigation of the nature of spacetime between the event horizon and the singularity of a non-rotating black hole by M.D. Kruskal and others shows clearly that any matter which enters such a black hole inevitably falls into the central singularity to be crushed beyond comprehension. Even if wormholes exist—which is very much open to debate—they would not offer a means of instantaneous transportation between one region of spacetime and another; the traveller would be destroyed in the singularity and it would be of little consequence to him if the crushed remnants of his constituent atoms were spewed up in some other part of spacetime!

However, we have been talking only about non-rotating black holes so far. Real objects in the Universe rotate, galaxies, stars and planets alike, and so we should expect black holes to rotate as well. The solutions of Einstein's field equations for a spinning black hole were published in 1963 by R.P. Kerr and among other features it showed that, in principle, it should be possible to enter a spinning black hole along a path which—without exceeding the speed of light at any point—avoids the central singularity and, apparently, emerges into "another universe". Again, some have argued, a spinning black hole may provide a gateway from our Universe into another or—perhaps—from one region of spacetime to another region of spacetime *in our own Universe* . Certainly, if we imagine a body falling into a spinning black hole and avoiding the singularity, since it is not permitted to re-emerge from that black hole (as to do so it would have to exceed the speed of light) in the vicinity of that black hole, it does not seem wholly unreasonable to suggest it may re-emerge "somewhere else".

If it were possible to use black holes as a means of instantaneous travel between different parts of our Universe, then again we could be faced with the most bizarre and paradoxical situations. For example, let us suppose that by charting a suitable route through a black hole (or a series of black holes) the crew of a starship could end up in any part of spacetime that they wished. They could then emerge at any convenient point in space *at any time*. It would be possible for them to devise a route which would allow them to return to the Earth prior to their departure. They could then be present to wish themselves well at their own departure—or even to persuade themselves not to bother setting out since they had already been on the journey! Even the simpler hypothetical case of using a black hole as an instantaneous bridge to another point in space raises the possibility of communicating information between

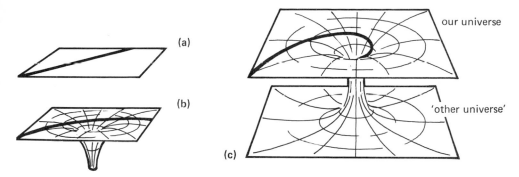

Fig. 19 The spacetime around a black hole. (*a*) Spacetime in the absence of matter may be represented by a flat surface. An object propelled along a flat surface will follow a straight-line path, just as a particle will do in the absence of a gravitational field. (*b*) If we suppose spacetime to be represented by a rubber sheet then if we place a massive body on that sheet, it will cause an indentation. The moving body from (*a*) will be deflected by this, just as a particle is deflected when passing a massive body. (*c*) The spacetime on the outside of a stationary black hole increases in curvature, then appears to open out again into another flat spacetime. It is as if a black hole connects our Universe to another universe; some have speculated that, instead, a black hole links different parts of our own spacetime. What, then, is the fate of a body that falls into a black hole?

two points in space faster than a light ray can cross that distance, and we have already looked at the paradoxes which would be raised by faster-than-light signalling! I must admit, though, that the idea of using a black hole for the instantaneous delivery of mail is very appealing.

These paradoxes would, of course, be avoided if a black hole provided a one-way link to a totally separate spacetime; i.e. to another universe. If by falling through a spinning (or, indeed, an electrically charged) black hole one disappeared forever from our Universe without prospect of return, then no problems of cause and effect would arise (one's sudden appearance in the "other universe" might greatly perturb the indigenous inhabitants, however)

On the other hand, if this possibility is admitted, there seems no good reason for excluding the possibility of matter suddenly appearing explosively in our Universe out of a "white hole"; the emergence of a white hole would be a wholly unpredictable event, and this unpredictability renders white holes undesirable in the eyes of most physicists.

As has been pointed out by N.D. Birrell and P.C.W. Davies, among others, the spacetime bridge which we have been discussing is an idealized concept which takes no account of the realistic physical situation of a black hole in the Universe. Taking into account the effects of matter surrounding the hole, and quantum effects, they conclude that it is most likely that the idealized "bridge" would be destroyed in the interior of a black hole. Like many other idealized concepts, the wormhole to another universe may not exist in practice. Nevertheless, this aspect of black holes remains an intriguing area which doubtless will see much more research activity and speculation, and which, potentially, could reveal further aspects of the mutable nature of time.

The Universe and Time

By this stage the reader could well be feeling that time is so ethereal a quantity that it must be impossible to make any definite temporal statements about the Universe. In fact the situation is not nearly so bad and it turns out that we can define a kind of *cosmic time* which characterizes the evolution of the Universe.

When we look around the Universe we find that it contains many billions of galaxies. Our Earth is a planet which travels around a star (the Sun) which is a member of one such galaxy—a system containing about one hundred billion stars. Although individual galaxies vary in size and structure, and although they tend to be arranged in groups or clusters, on the large scale, the distribution of galaxies seems to be isotropic; i.e. we see the same large-scale distribution of galaxies in every direction we look.

Another fundamental observation is that, with the exception of our immediate neighbour galaxies, making up the Local Group of galaxies, all the galaxies show redshifts in their spectra. This we interpret as meaning that the galaxies are receding from us, the redshift being an indication of the velocity at which the galaxies are moving away. It was shown in the nineteen-twenties by E.E. Hubble that the speed at which a given galaxy is receding is directly proportional to its distance: the more distant a galaxy, the faster it is seen to be moving. The velocity of a galaxy (V) is related to its distance (D) by the simple formula, $V = H \times D$, where H is a constant known as Hubble's constant.

Although at first glance these observations seem to imply that all the galaxies are receding from us in particular, and that we, therefore, must be at the centre of the Universe, *this is most unlikely to be so*. The observed expansion of the Universe appears to be perfectly symmetrical, so that whichever galaxy you observed from you would see the same general picture—all the galaxies would appear to be receding from you in particular. (It is, of course, possible that we are at "the centre", but ever since Copernicus dethroned the Earth from its central position in cosmology in the sixteenth century, we have tended to regard with suspicion any observation or theory which appears to assign to the Earth a unique and privileged position in the scheme of things.)

Because of the observed large-scale uniformity and isotropy of the matter in the Universe (and because it makes the mathematics easier), modern theories of cosmology are worked out on the assumption that matter is uniformly smeared out through the Universe, constituting a "sub-stratum" which is expanding. An observer who is at rest with respect to the "smeared-out" matter in his vicinity (i.e., with respect to the sub-stratum) is called a *fundamental observer*. He is in a privileged position because he is sharing in the overall uniform expansion of the Universe. Since the galaxies are participating in this expansion, it is reasonable to suppose that an observer located in a galaxy is, to a good approximation, a fundamental observer. The *cosmological principle*, central to most theories today, is that the Universe as seen by fundamental observers is homogeneous (i.e., every fundamental observer sees the same broad picture of the Universe as time goes by) and isotropic (the Universe looks the same to each fundamental observer in every direction in which he looks). There is, then, no unique centre to the Universe, and no discernible "edge"; if there were a centre and an edge, we should expect to see a concentration of matter in one direction (towards the centre) and a thinning out of matter in the other (towards the edge).

It is difficult for us to visualize a Universe which has no centre and no edge. However, the curved spacetime of general relativity lends itself to the possibility that we live in a Universe which is finite in extent yet unbounded, and which is expanding from a common origin a finite time ago. A common analogy for a finite yet unbounded Universe is to represent the Universe by the *surface* of a balloon. A flat, two-dimensional creature on the surface of a sphere—who is not aware of the existence of the vertical direction—would have no means of visualizing the sphere on which he sat. Yet, by experiment, he could find out that his "Universe" was finite but unbounded. If he set off in one particular direction, and kept going, he would eventually return to his starting point; he would have circumnavigated his Universe without ever having come to an edge. Indeed there are measurements he could make which would confirm his mathematical model that his Universe is a sphere even though he could not *visualize* what a "sphere" looks like (a sphere exists in three dimensions, but he is aware of only two). In an analogous way, our

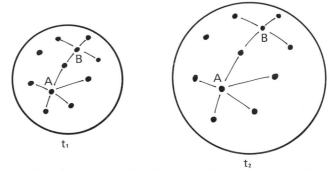

Fig. 20 Analogy for the expanding Universe. If we represent the Universe by the *surface* of a balloon, and the galaxies by dots upon that surface, then we symbolize the expansion of the Universe by the inflation of the balloon. Thus the separation of all the galaxies increases, and the *scale* of the Universe becomes greater. If an observer in galaxy A measures the distances to the galaxies within range of his telescope at epoch t_1 and then at later epoch t_2, it will appear to him as if he were at the centre of the expansion, since all the other galaxies will have moved away from him. However, an observer in galaxy B—or any of the other galaxies—will conclude that *he* is at the centre. In fact, no galaxy has any better claim to being "at the centre" than any other: indeed, such a universe does not *have* a centre.

Universe may be curved in such a way as to be finite and unbounded, even though we cannot really visualize the situation.

If we represent the galaxies by dots painted on the surface of a balloon (Fig. 20) then each galaxy will "see" the same general picture of the Universe (bear in mind that the "inside" and "outside" of the balloon do not constitute part of the space of our Universe). If we expand the balloon, then the separation between the "galaxies" will increase symmetrically, each galaxy moving away from every other galaxy. Each galaxy will see all the others move away from itself in particular, but it is quite clear that every galaxy "sees" the same picture. No one galaxy can claim to be the centre of this expansion. What is happening is that the space between the galaxies is increasing; the *scale* of the Universe is changing. In effect, the space between the galaxies is growing, but the galaxies are maintaining the same configuration in relation to each other. The galaxies are not moving through space, but are remaining at rest in the expanding space of the Universe. Observers on such galaxies are the "fundamental observers" to which we referred earlier, remaining at rest relative to the general expansion of the sub-stratum (the surface of the balloon) in their vicinity.

If the Universe is expanding, as it seems to be, then it is reasonable to suppose that at some time in the past all the galaxies must have been adjacent to each other, and if we trace back further, all the matter in the Universe must have been concentrated together in an intensely hot fireball of matter and radiation. It is currently fashionable to accept the notion that the Universe originated in a hot Big Bang (we can deduce what must have been the temperature of this "Big Bang" by examining what would happen if we reversed the expansion of the Universe and allowed all the galaxies to fall together), an explosive event which hurled matter and radiation in every direction—the recession of the galaxies being due to the violence of the explosion.

However, the Universe did *not* begin as the explosion of a super-dense concentration of matter into a previously existing empty space. Instead, general relativity suggests that space has expanded with the matter; in effect both space and time originated with the Big Bang. The Big Bang was a singularity in spacetime similar to the singularity which is assumed to exist in the centre of a black hole. The known laws of physics can, with a degree of confidence that may not be wholly justified, be used to describe event as far back as one millionth of a second after the "beginning", but we have no knowledge of what happened prior to that. It is pointless to ask "What happened before the Big Bang?" because it looks as if space and time in the sense in which we use these terms simply did not exist before that instant. The Big Bang was the origin of time.

If we return to a consideration of the fundamental observers, stationary relative to the expanding framework of the Universe, and shown on our simple model by the dots on the surface of an expanding balloon, we note that each observer sees the same general view of the Universe as every other observer at every stage of the expansion of the Universe. In other words, every observer sees the same succession of states through which the Universe passes as it evolves; all fundamental observers agree on the order of the successive states of the Universe. Each observer can label time by the succession of states through which the Universe passes, and since each observer sees the *same* succession of states, then their clocks may be synchronized to give one common universal time, known as *cosmic time*. In relation to cosmic time, all events have a unique time order. Discrepancies in the order of events which can arise in special relativity crop up because we are dealing with observers who are moving relative to the fundamental observers in their vicinity. By no stretch of the imagination could relativistic space travellers *en route* from one galaxy to another be considered as fundamental observers: they would clearly be moving relative to the sub-stratum. But perhaps observers like us, located in a galaxy, might be considered as idealized fundamental observers.

In order that there may exist a common cosmic time the Universe must be highly isotropic—to a fundamental observer the Universe should look the same in all directions. Recently it has been shown that, in one respect at least, the Universe is very isotropic indeed. If, as current theory and observation seem to suggest, the Universe originated in a hot fireball of matter and radiation, then the whole Universe should be filled with a weak background of radiation left over from the Big Bang. Because of the expansion which has taken place since the early dense state of the Universe, this radiation should now be detectable in the form of microwave radiation. Radiation of this kind was first detected in 1965 by A.A. Penzias and R.W. Wilson, who were awarded the 1978 Nobel Prize in Physics for this achievement. Subsequent observations of its intensity in different directions have shown that the cosmic background radiation is isotropic to within one part in 3,000. This is a very high degree of isotropy, and leads us to believe that all fundamental observers will see the same picture if they examine the background radiation. This in turn points strongly to the validity of the concept of cosmic time.

To what extent may we regard ourselves as being fundamental observers? Of course we are moving due to the rotation of the Earth and to its motion round the Sun. Likewise, the Sun moves around the galactic centre, and the Galaxy itself is moving relative to the neighbouring galaxies which make up the Local Group. These velocities are small compared to the velocity of light. Does our Galaxy (or the Local Group of galaxies) have any other *peculiar* (i.e., individual) motion relative to the sub-stratum? The cosmic background radiation provides a means of determining this velocity—in effect, it provides a new kind of "ether" against which to measure our velocity relative to the bulk distribution of matter and radiation in the Universe. If the Earth is moving through the background radiation, then the radiation ahead of of us will be "blueshifted" (i.e., by the Doppler effect, the approaching wavecrests will be "squashed up" resulting in the observed wavelength being shorter than the average wavelength of the background radiation) while the radiation approaching from behind will be redshifted (i.e., it will arrive with a longer wavelength). Observations of this kind show that the Galaxy is moving relative to the background radiation with a velocity which may be as high as 600 kilometres per second. The value is surprisingly high, and one suggestion is that our Galaxy is being accelerated by the local "supercluster" of galaxies. However, the fact remains that this velocity is only one five-hundredth of the speed of light, so that we are not too far removed from being ideal fundamental observers: our view of the Universe should not differ greatly from the view of a hypothetical fundamental observer.

In regarding the cosmic background radiation as a kind of "ether", we are not reverting to Maxwell's ether or to the Newtonian idea of absolute space. The motion of the Earth, or of the Galaxy, relative to the background radiation is not motion relative to some fixed absolute frame of reference, but it *is* motion relative to what is, perhaps, the most natural reference frame in the Universe, the expanding coordinate system in which the galaxies as a whole are at rest.

A consideration of the expanding Universe, then, leads us to the notion that time began with the Big Bang and that, if we imagine a set of fundamental observers who are at rest with respect to the expanding coordinate system of the Universe, we can devise a system of cosmic time which will allow these observers to assign a unique date to events in the Universe. This is not "absolute time" in the Newtonian sense—which "flows equably without relation to anything external"—but it is, perhaps, the next best thing! Anomalies of time measurement—the length of time intervals and the order of events—arise when we consider the viewpoints of observers moving relative to the fundamental observers. There is no doubt that if a high-speed space mission were undertaken whereby one member of a pair of twins stayed at home and the other went out and back at relativistic speed, the astronaut twin would return to find that he had aged less than his Earth-bound brother. The existence of a cosmic time makes no difference to that, for we are not here dealing with fundamental observers. Likewise, the curious temporal effects associated with, for example, black holes, are unaffected by the possible existence of a cosmic time, for again we are dealing with observers who are violently accelerated compared to fundamental observers in their locality. Time is still mutable, it is just that cosmic time allows us to make sensible statements about the evolution of the Universe as a whole.

What of the future? If space and time originated with the Big Bang, will they ever come to an end? A number of possibilities present themselves. If there is sufficient matter in the Universe, the combined gravitational attraction of all the matter will be sufficient eventually to halt the recession of the galaxies, but if the quantity of matter is insufficient then the expansion will continue forever. Admittedly, the expansion velocity will slow down, but it will never reduce to zero. An analogy can be drawn with the idea of escape velocity: if the galaxies have sufficient velocity (or, strictly, kinetic energy) they will continue to move away; if they have just sufficient velocity (the "escape velocity") they will move to an infinite distance with a velocity which reduces to zero; if they have insufficient velocity, then they will recede so far and fall back together again. The present observational evidence seems to show that there is not nearly enough matter in the Universe to halt the expansion, and it looks as if the Universe will continue to expand without limit: time will go on forever. There is still some doubt about the matter, so let us consider what would happen if the Universe expanded so far and then began to collapse. Ultimately, all the matter of the Universe would fall together at a point, a new spacetime singularity where the structure of spacetime as we know it would be destroyed. In that case, perhaps space and time would

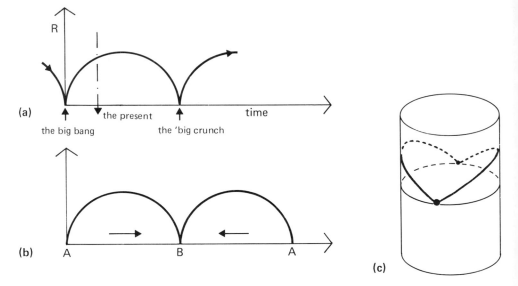

Fig. 21   Closed time and the oscillating Universe. (a) One theory suggests that the Universe expands to a certain size, then collapses again, in a regular cyclic way: the present cycle need not be the first or the last—indeed, the Universe may continue to oscillate in this way for ever (although the present observational evidence does not support this view). (b) A possibility suggested by P.C.W. Davies is that time is closed like a loop in such a way that temporally asymmetric processes, such as the increase of entropy, act in opposite directions (as shown by the arrows) in successive cycles. The Universe expands from a Big Bang at A until it eventually collapses to a "Big Crunch" at B; in the next cycle the condition of the Universe returns to its starting point, A. In effect, it is as if the diagram of the two cycles of the oscillating Universe were wrapped around a cylinder, as in (c), so that the two points A coincide.

come to an end in the "Big Crunch", just as space and time began in the Big Bang.

There is a theory which suggests that if the Universe collapses on itself, it will "bounce" in a new "Big Bang" and will then enter a new phase of expansion. Indeed, the present cycle of the Universe may not be the first, or the last; we may live in an oscillating Universe which expands and contracts in a periodic way. Such a theory has a certain appeal; there is no unique beginning to it all, and the Universe always remains finite in its spatial extent. At the end of each cycle, the contents of the Universe are completely reprocessed and the new cycle starts completely afresh; it is even suggested by some that the laws of nature may be changed in successive cycles so that only some cycles of the Universe will be suitable for the emergence of life and it is only by chance that in the present case there are cosmologists around to wonder about it all. However, it is no more than an article of faith to suppose that such a rebound could occur; if the Universe did collapse upon itself that might spell the end of it—there might well be no subsequent* cycle.

To take matters further, might not time, or temporal processes, run backwards in the collapsing phase of the cycle, or perhaps one complete cycle of the Universe is followed by another in which time somehow "runs backwards"? Before considering such possibilities, let us examine anew the whole question of the "flow" of time.

The Arrow of Time

At the beginning of this chapter we saw that the idea of time "flowing" was quite unsatisfactory. For time to flow, its rate would have to be measured against some more fundamental kind of time or, alternatively, time would have to flow relative to itself—a logical absurdity. Indeed, if we accept a four-dimensional spacetime description of the Universe then spacetime just exists: it cannot flow. Physicists attempt to circumvent the problem by abandoning altogether the idea that time flows and replacing it with the notion that time, or temporal processes, are *asymmetric.*†

---

* All discussion of time is hampered by semantic difficulties. The use of the word "subsequent" here appears to imply that time goes on independently after the Big Crunch, whereas it seems likely that space and time cease at that point.

†The general question of time asymmetry is discussed in, for example, *The Physics of Time Asymmetry*, by P.C.W. Davies, Surrey University Press, 1974, and at a more popular level by the same author in *Space and Time in the Modern Universe*, Cambridge University Press, 1977.

Nevertheless, in everyday existence we are very conscious of the impression that time flows past us, that the future becomes the present, and that present events become the past. The whole idea of "future", "present" and "past" is conditioned by our day-to-day impression of time's flow. There is a profound difference between the past and the future in that we can remember the past (and we can see the cause-and-effect relationship between past and present events) but we cannot know the future. We may attempt to *predict* the future, but we cannot make definite statements about future events. Likewise we cannot imagine that future events can influence present events. Certainly, then, there is an asymmetry about the nature of time. Accepting the *caveat* that it is misleading in a strict sense to talk about the flow of time or of the moving present moment, it is convenient and easy to visualize time as having a direction, and to talk about the *arrow of time* as being the direction in which the present moment flows, from the past into the future.

All the macroscopic phenomena of our world show that the arrow of time points in the same direction—towards the future. The unidirectional nature of time shows up in a variety of ways. For example, biological evolution appears to be irreversible, having proceeded on Earth as a result of a long chain of chance mutations towards ever more complex states, and, in the process, modifying the environment of the Earth. With the increase in complexity of the organisms and their environment the chance of precisely the same conditions being restored to allow, say, the dinosaurs to reappear on Earth must be incalculably small. Biological evolution, then, is a one-way process.

Or again, a star originates out of the collapse of a cloud of gas in space and then, when its interior becomes sufficiently hot, it generates energy by means of nuclear transformations which convert hydrogen to helium, then helium to carbon, and so on. Finally, the star runs out of fuel and ends up as a dense white dwarf, an even denser neutron star, or—perhaps—a black hole. We do not see the reverse process happening: a white dwarf cannot become a main sequence star and then decompose into a primitive cloud of interstellar gas. The observed expansion of the Universe, which we have already discussed, is another example of a one-way process; but here we are on less certain ground, for, although the present evidence suggests that the expansion should continue without limit, the possibility of a future contracting phase cannot be completely excluded. E.A. Milne was one of the first to suggest that the expansion of the Universe defines the direction of time's arrow (would such a supposition imply the reversal of time's arrow during a contracting phase?).

The Second Law of Thermodynamics, as interpreted by L. Boltzmann in 1866, suggests that any closed system will tend towards the state of greatest disorder: the *entropy* (which is a measure of disorder) of a closed system must always tend to increase. For example, if we have a pot of coffee and a pot of milk, then we have some degree of order in that coffee and milk are separated from each other. If we now pour some of each into a cup and stir the mixture we end up with white coffee, and there is no way that this chaotic mixture is suddenly going to separate itself into its two basic constituents. If the Second Law were applied to the Universe as a whole, then in time the entire Universe should tend towards a state of maximum entropy: the stars should continue to radiate heat, light and energy until the entire Universe reaches an amorphous state in which no change is possible. This hypothetical state is termed the *heat death of the Universe*, and it may indeed be that, if the Universe continues to expand without limit, this is the state towards which it is heading.

When we turn to the basic laws of nature, however, we can find no evidence of the one-way nature of time. The laws of nature would be equally reasonable whether time were flowing "forwards" or "backwards". For example, consider a ball which falls to the ground and bounces back to its starting point (an ideal ball!). If the direction of time were reversed, the ball would still be seen to fall down and return to its starting point. The law of gravity is certainly symmetrical with respect to time. A comet moves round the Sun in a particular direction, and, if time were reversed, it would pursue the same orbit but in the opposite direction. If we were to observe this comet, it would still be following a path entirely consistent with the law of gravity. There is nothing to specify the direction in which time must flow. The same is true for the laws of electromagnetism and those governing the strong nuclear reactions in the nuclei of atoms. Atomic particles are quite indifferent to the direction of time's arrow, and nothing can be deduced from their study which suggests why time should "flow" in one direction rather than another.

This utter indifference of atomic particles and physical laws to the direction of time's arrow is known as *time-reversal symmetry* and is denoted by T. Only in one specific situation involving the weak nuclear interaction has any doubt been cast on the validity of T.

Nuclear physicists, in their attempts to unravel the complex world of so-called "elementary" particles, are concerned to identify symmetries, patterns of behaviour which allow some sense to be extracted from a confusing picture. A hallowed symmetry at present is CPT invariance. C denotes charge conjugation (the process of changing a particle into its antiparticle) and P denotes parity (the symmetry between "right-handed" and "left-handed" aspects of nature). Although both C and P are violated in certain nuclear reactions it is strongly felt that the combined operation CPT should be inviolate; i.e., that it should be preserved in all particle reactions. If it is not, then a cornerstone of modern physics will have crumbled. In 1964, J. Christenson, J. Cronin, V. Fitch and R. Turlay of Princeton University showed there are particles known as K-mesons which decay in particular ways to form particles called pions in a way that clearly violates CP, the operations C and P taken together. If CPT is to be preserved, this can be achieved in this reaction only if T also is violated. The violation of CP "cancels out" the violation of T so that the operation CPT is preserved. It looks, then, as if the decay of these particles shows that there is at least one process at the subatomic level which does violate time-reversal symmetry. That example apart, there is nothing in microscopic physics to suggest that time must flow one way or the other.

We have, then, a dilemma. On the one hand, large-scale phenomena in the Universe all point to time flowing uniquely in one direction, but with one minor exception subatomic particles and the fundamental laws of nature are quite indifferent to the direction of time's arrow. Although David Layzer has recently developed a theory which suggests that time's arrow was determined by the initial conditions of the universe, it must be admitted that there is no general agreement at present as to why time's arrow points in one direction!

Could, then, time ever flow backwards in the Universe on a large scale? It might be rather fun if it did. Elderly people would evolve towards childhood, demolished buildings would rise from the dust to assume their original pristine states, ripples would converge on pebbles which would then leap into the hands of the people who once threw them in the water, and so on. Life would be like a film running backwards, if the direction of time truly were reversed. There is no evidence to support the possibility of this happening, and there is little likelihood of temporal processes being reversed if the Universe ceased to expand and entered a contracting phase.

A further possibility has been debated by P.C.W. Davies who considered the possibility of closed time, where time is closed like a loop (a possibility which we touched on at the beginning of this chapter). He considered a Universe which expands to a maximum volume then collapses ultimately to enter a new cycle of expansion and contraction as before. The central point considered was that the direction of temporally asymmetric processes (such as the increase of entropy) was reversed in the second cycle, so that at the end of the second cycle, the Universe had returned to its starting point; i.e. to the beginning of the first cycle (Fig. 21). In a crude sense, this implies the reversal of time processes at the end of each cycle. An interesting suggestion made by Davies is that starlight emitted during one cycle will appear as background radiation in the other. Food for thought indeed!

Coda

In conclusion, what then is the status of time? The absolute time of Newton has been swept away to be replaced by the mutable time of relativity. In special relativity we see that the time interval between events will be assigned different values by observers in uniform relative motion but, so long as the events are causally related, or could be connected by a signal which does not exceed the speed of light, then at least the order in which events occur is preserved. If, however, the separation in space between two events is such that they cannot be connected by a ray of light, then different observers may see even the order of the events reversed.

Space and time, it seems, cannot be regarded as separate entities. Instead, we have to follow the advice of Minkowski and treat the three dimensions of space and the dimension of time as a four-dimensional structure which we call spacetime. The effect of matter is to distort spacetime, so affecting the paths pursued by material bodies, and by rays of light in the vicinity of that matter. From this point of view, gravitation is seen not as a force acting instantaneously across space from one body to another but as the curvature of spacetime which influences the motion of the particles which lie within it. From the general theory of relativity we find that the rate at which time passes depends upon the strength of the gravitational field in which a clock or an observer is placed.

Time travel is possible, but only in the one-way sense. An astronaut making a long journey to a distant star—and back—at a high proportion of the speed of light *will* return to find that many more years have elapsed on Earth than have passed in his spacecraft.

Can rotating black holes be used to provide instantaneous travel to other parts of the Universe, or even to other universes? The mathematics might seem to indicate it. Here a "black hole probe" is approaching a rotating hole in a system like that of Cygnus X-1, where the black hole is revolving with a hot massive star. The long needle-shaped part of the probe is for use in measuring tidal effects as it approaches the hole; the tiny payload capsule at the tip contains superdense arte-facts which, it is hoped, will survive the journey into another universe as Man's first messages. Of course, first of all one has to travel to one's black hole . . .

But the journey he has made is into the future of the Earth. There is no way (provided that the velocity of light cannot be exceeded) that a journey may be undertaken into the past. The relativistic traveller cannot return to the era of his contemporaries back on Earth. In a similar way, a forward trip could be made to the Earth's future by an astronaut who spent some time in an intense gravitational field but, again, there is no way back. Strange things happen to time at high velocities and in intense gravitational fields, of that there is no doubt.

Nevertheless, the Universe appears to be so uniform and isotro-pic that we can devise a system of cosmic time which allows us to date absolutely the sequence of events which have taken place since the formation of the Universe. Newton's absolute time has gone, but temporal ordering in the Universe is not a free-for-all.

At the subatomic level, and so far as the basic laws of nature are concerned (with one minor exception), there is no reason why time should flow in one direction rather than the other, but on the large scale, the processes taking place in the Universe all seem to indicate that time has a unique direction. Why time exhibits this obvious asymmetry, what would happen if one fell into or "through" a spinning black hole, what is it that governs the time order of our perception of events? These are but a few of the questions which remain to be answered. Time is fundamental to life, and will remain a key debating issue for philosophers, physicists and laymen alike. This chapter has come full cycle. We may measure time intervals as precisely as we wish, but we seem to be not very much further advanced in understanding what time *is*!

IKMN

6 Measuring Time Past

Early Beliefs

Our understanding of time has grown in step with the discovery and development of methods of time measurement over periods beyond the range of human experience. As we have seen in earlier chapters, mankind has, for many centuries, been familiar with ideas concerning the division of the day into time units and the division of the year into seasons or months; but ideas of change and development or evolution in the natural world were slow to find acceptance, since the population was generally unskilled in the detailed observation of minor environmental variations. Until about a century ago, there were no accurate scales available for the measurement or interpretation of time past, although Man was capable of recording regular or recurring events on a scale of years or even human generations. There was no absolute timescale; everything was relative, and all references to time past were made in terms of human experience.

In most religions and cultures time was thought of as repetitive or cyclic. On the other hand, the Greek philosophers of the period 500-300BC were familiar with the idea that the surface of the Earth was not static but was changing continuously. They thought of time as linear, and they speculated about the evolution of landscapes and the evolution of life. Later, as described in Chapter 1, the rise of Judaism and Christianity emphasized this new view of time as linear and irreversible. As in other religions, there were the concepts of Creation (the beginning of the world and the beginning of time) and Destruction (the end of the world and the end of time), but the Greek concept of eternity was given great emphasis. Eternity was a sort of supertime, understood and created by God alone. In eternity, existence had meaning, but time had not. The idea of eternity has exercised theologians and philosophers for many centuries, and much of the Bible is concerned with attempts to explain its meaning. Perhaps "eternity" is simply a word used to describe time on a scale which is beyond human comprehension: time before the advent of Man and also time beyond the end of Man and the end of planet Earth. Perhaps the concept of eternity evolved as the result of a corporate insight that there is an absolute timescale for the history of the Universe—a timescale of far greater significance than any relative timescale based upon human experience.

The linear timescale used in geology and many other sciences may be familiar to us, but it is worth remembering that the earliest linear timescales were invented less than 200 years ago. The Greek philosophers may have been familiar with the idea of environmental change and the evolution of life, but they had no conception of either a relative or an absolute timescale. Aristotle (384-322BC) summarized their views as follows: "The distribution of land and sea in particular regions does not endure throughout all time, but it becomes sea in those parts where it was land, and again it becomes land where it was sea . . . As time never fails, and the Universe is eternal, neither the Tanais nor the Nile can have flowed for ever . . . So also of all other rivers; they spring up, and they perish . . ." Aristotle knew of the importance of water in shaping the face of the land, but he gave no estimates of the amount of time involved in the processes of erosion and the deposition from the water of material it had eroded. He knew that the Nile Delta had been built up by the slow deposition of sediments from the river, but he had no scale for estimating the *rate* of deposition or the passage of time since the delta began to form. The Roman writers Strabo, Seneca and Pliny made excellent observations on the changing environment and the nature of fossils, but they too were handicapped by the lack of a timescale. The same problem confronted the Arabian scholars such as Avicenna (980-1037). Leonardo da Vinci speculated on a number of geological and geomorphological features and gave an extremely accurate explanation of fossils, but again he suffered from a lack of a reasonable time framework for his observations.

With the wide acceptance of Christianity in western Europe, the Bible took on an authoritative rôle in discussions of a geological and biological nature. As we saw in Chapter 1, a Biblical timescale was evolved, and some scholars became involved in the calculation of dates for each of the Biblical "ages". The book of Genesis was to be interpreted literally. The new churches established by the Protestant Reformation were in many instances more fanatical than the Catholic Church in regarding the Bible, literally translated, as the guide to all thinking on natural history. During the sixteenth century ludicrous ideas concerning the surface of the Earth and the nature of life abounded—but we must not forget that original scientific enquiry was not only officially discouraged but also extremely dangerous; anyone who disputed the authority of the Bible was likely to lose his life. Two hundred years after Leonardo da Vinci, science had reverted to a stage of utter naïvety, largely because of the influence of the Christian Church.

Catastrophism

In the eighteenth century there was still a widespread belief that change in the environment was achieved through a series of disconnected catastrophic events. There was a strong Biblical basis for this belief. The most important geomorphological events were of course the Creation and the End of the World. Next in importance was Noah's Flood; and in keeping with the emotional climate of the times other major natural happenings such as earthquakes, avalanches, hurricanes and volcanic eruptions were looked upon, in most religions, as the wilful acts of deities. Even today the primitive tradition remains with us—natural disasters are almost invariably still referred to as "Acts of God", in spite of the fact that we normally look upon God as a benevolent, rather than a vengeful, deity.

A great deal of scientific attention was concentrated upon Noah's Flood, and this single event was used to explain the shaping of the land surface, the layering of sedimentary rocks, the occurrence of fossils and many other geological phenomena. The Neptunists, led by the German Abraham Werner, believed that the majority of the Earth's rocks had been precipitated in a universal sea, and complicated chemical analyses were presented in support of a three-fold division of rocks based upon their order of precipitation.

During the late eighteenth century the Diluvialists began to realise that not all natural phenomena could be explained by reference to *one* great flood, and more realistic interpretations of geological time began to appear. William Buckland, a theologian who became a famous professor of geology at Oxford, dealt with some of the problems of stratigraphy by adding two more worldwide floods to Noah's Flood. Georges Cuvier postulated a whole series of floods caused by sudden crustal subsidence. These floods intermittently wiped out animal and plant life in Europe, and after each flood sudden uplift of the Earth's crust presented a fresh land

A portrayal of Noah's Flood by Francis Danby (1793–1861). The Flood was thought of, particularly in the seventeenth and eighteenth centuries, as one of the main events of geological time.

surface which could be colonized by organisms from other areas.

Gradually the original simple ideas of catastrophism were giving way to more complicated theories, and these theories evolved further as scientists were forced to recognize the great complexity of the natural world and the great span of geological time.

A Time of Enlightenment

A realistic sense of time came to the Earth sciences through the gradual acceptance of the "principle of uniformitarianism". This principle, published in 1788 by James Hutton, stated that the formation of ancient rocks could be explained without invoking any processes other than those which could be directly observed. Further, Hutton believed that all geological processes that operated in the past could also be observed at the present time. The history of the Earth could be explained without recourse to catastrophes, divine intervention, or processes that could not be directly measured or tested.

Over a century before Hutton, a Dane called Nicolaus Steno had recognized that sedimentary strata are laid down layer upon layer, that younger beds are laid down on top of older beds, and that strata are initially deposited in horizontal layers. These principles, published far ahead of their time and originally ridiculed, could

William Buckland (1784–1856), the British geologist, equipped to explore a glacier.

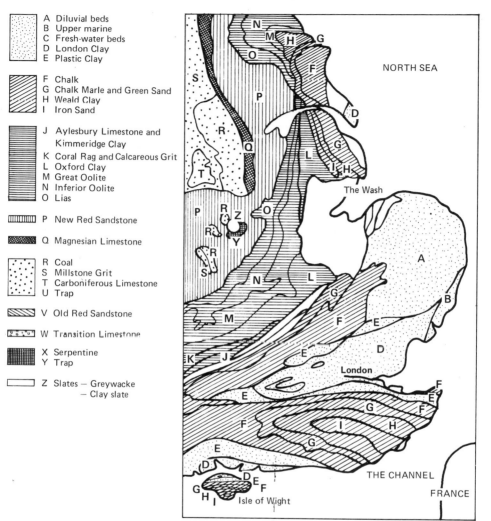

A Diluvial beds
B Upper marine
C Fresh-water beds
D London Clay
E Plastic Clay

F Chalk
G Chalk Marle and Green Sand
H Weald Clay
I Iron Sand

J Aylesbury Limestone and
   Kimmeridge Clay
K Coral Rag and Calcareous Grit
L Oxford Clay
M Great Oolite
N Inferior Oolite
O Lias

P New Red Sandstone

Q Magnesian Limestone

R Coal
S Millstone Grit
T Carboniferous Limestone
U Trap

V Old Red Sandstone

W Transition Limestone

X Serpentine
Y Trap

Z Slates — Greywacke
            — Clay slate

NORTH SEA

The Wash

London

THE CHANNEL

FRANCE

Isle of Wight

The British geologist William Smith (1769–1839) is regarded as the father of stratigraphy; he realized that the succession of the rocks provided a guide to the nature of times past. Here is a diagram based upon his 1815 map of the south-eastern part of the UK. For its time, the map is remarkably accurate.

now be interpreted in the light of Hutton's ideas, and modern geology was born.

Its early faltering steps were helped by an English surveyor called William Smith, who observed that "the same strata were found always in the same order and contained the same peculiar [i.e., characteristic] fossils". These observations gave rise to the important geological principle which emphasizes the succession of fossil assemblages through geological time and the correlation of rock sequences based on fossils. Smith's geological map of

England, Wales and part of Scotland, published in 1815, is one of the great geological source documents, and his geological cross-sections and columns were also of pioneering significance.

The new ideas of geology and the new appreciation of the span and continuity of geological time were helped along by two books which made a profound impact. These were *Illustrations of the Huttonian Theory of the Earth*, by John Playfair, published in 1802, and *Principles of Geology*, completed by Charles Lyell in 1833. The former made the ideas of Hutton widely available in simple yet scientific language, and the latter developed the principle of uniformitarianism as the theme of a highly successful textbook. By the mid-1830s the catastrophists, the fundamentalist theologians, the Diluvialists and the Neptunists were all retreating in disarray, and in spite of occasional rearguard actions their battles were already lost.

After 1833 the scientific study and subdivision of geological time became a respectable branch of geology, and "historical geology" has remained popular to this day. At the same time, the new generation of geologists accepted that *gradual change* and *evolution* were key concepts in their subject. Research work commenced into the ways in which natural processes changed the environment, and studies of the work of rivers, glaciers and other agents of erosion began to appear.

In the early years of the nineteenth century, a number of observers such as Jens Esmark, Ignaz Venetz-Sitten and Jean de Charpentier realised that the glaciers of Scandinavia and the Alps had once been more extensive, and proposals were made that the greater part of Europe north of the Alps had been covered by glacier ice. The leading proponent of these ideas was Louis Agassiz, and his enthusiastic campaign for the acceptance of the "glacial theory" was supported by well known geologists such as William Buckland and Archibald Geikie. By about 1860 it was widely accepted that both northern Europe and a large part of North America had been overridden by ice sheets. The theory of ice ages was also gaining acceptance. The idea that the world suffered intermittent periods of severe cooling and glaciation was of great importance, for it brought home to many nineteenth-century Earth scientists the instability of the global climate. The large-scale climatic oscillations required for ice ages were interpreted by some as natural catastrophes, while others looked on them as short-lived departures from the "normal" environment.

But, however they were interpreted, it became apparent that the

Ice on the face of the land—a highland icefield and glaciers, seen to the north of Surprise Fjord, Axel Heiberg Island. The rise of the Glacial Theory marked a great advance in the understanding of geological time and the appreciation of changing climates on a global scale.

idea of a "steady-state" Earth was no longer valid. The view of geological time proposed by Hutton and Lyell was one of gradual change, with erosion and deposition on the one hand and crustal uplift and subsidence on the other hand maintaining an overall global equilibrium. In the new vision of Earth history time was not a cycle but an arrow: the sequence of events since the beginning of geological time was unrepeatable and irreversible, and ice ages could simply be interpreted as wobbles or fluctuations which disturbed the course of the arrow.

The British naturalist Charles Darwin (1809–1882). Through his theory of evolution by natural selection, Darwin argued forcefully in favour of an immense span of geological time.

In the life sciences, as in geology, the final abandonment of Biblical dogma by the scientific establishment followed the publication of Charles Darwin's *The Origin of Species* in 1859. Darwin was not the first scientist to propose the extinction and evolution of species, and indeed in 1705 Robert Hooke had suggested the use of fossils as chronological indices, the extinction and development of species, specific variation and progression due to changed environmental conditions, and climatic changes being inferred from fossils. But Darwin, much influenced by the uniformitarian views of Hutton and Lyell, presented his theory of evolution by natural selection in such a careful and comprehensive way that its acceptance was inevitable. A furious controversy concerning the evolution of Man clouded the broader issues involved, but Darwin provided a dynamic rather than a static view of the history and development of life on planet Earth, and this view was closely similar to the new geological ideas of dynamic and evolving global environments and landscapes. Further, biologists and archaeologists were now also forced to accept the notion of a long and gradual Earth history in which natural catastrophes such as Noah's Flood were not necessary for the explanation of the natural world.

By the year 1900 most of the key concepts associated with geological time were well established. The gradual evolution of the land surface, the gradual accumulation of sediments, and the gradual evolution of life forms were parts of conventional scientific belief. It was also believed that *change* was an essential characteristic of time; the progression of life forms, the progression of climates and the progression of landscapes all demanded a linear view of time in which every unique set of circumstances arose from a unique set of preceding circumstances.

It was already accepted that tectonic forces and periods of mountain-building were of great importance in the explanation of Earth history and the history of life, but before 1915 it had not been widely accepted that forces originating in the centre of the Earth were capable of altering the positions of the continental landmasses themselves. Climatic change was easier to accept than the migration of continents, and for this reason Alfred Wegener's theory of continental drift was at first ridiculed and largely rejected by the scientific establishment. This rejection occurred in spite of the very forceful arguments presented by Wegener from a wide variety of different fields, and it was not until the 1960s that the theory gained its rightful place as one of the fundamentals of modern geology.

Now our vision of a mobile and dynamic Earth is a comprehensive one; throughout geological time we have to interpret natural phenomena in the context of changing climates, changing environ-

Changing Views of Geological Time, from the Seventeenth Century to the Present Day

Year	Authority	Line of Reasoning	Estimated Age
1664	Archbishop Ussher	Biblical evidence	4004BC
1800	Comte de Buffon	Cooling of iron	75,000 years BP
1854	Herman von Helmholtz	Sun's luminosity	20–40 million years BP
1897	Lord Kelvin	Cooling of Sun	20–40 million years BP
1908	Joly	Salt in oceans	80 million years BP
1911	Boltwood and Holmes	Radiometric dating	c.2,000 million years BP
1970	Various	Meteorite and Moon rock dating	c.4,600 million years BP

ments, changing landscapes, and changing continental arrangements. Time may be linear, but the line has not been straight.

Towards a Timescale

The key rôle of time in the science of geology is apparent if one looks back at the development of the subject. Over the past 200 years or so there has been halting progress towards the establishment of a reliable timescale. At first the subdivision of geological time was somewhat erratic, and the various ages of rocks were given relative to one another. Because there were no means of absolute dating, the early geological timescales were thus *relative timescales*. The eighteenth-century subdivision of rocks by the Italian Arduino was a simple one:

Primitive: crystalline rocks in the cores
of mountains;
Secondary: sedimentary rocks;
Tertiary: unconsolidated sediments; and
Volcanics: extrusive igneous rocks.

Other works used the terms Primary, Secondary and Tertiary for various sorts of rock, and although these terms are still used occasionally by geologists they originally had nothing to do with rock *ages*.

The establishment of a timescale required a much more sophisticated terminology, and in a burst of activity, following the lead of

The Earth through time. These maps show the changing disposition of continents and oceans at widely spaced geological intervals: (a) Precambrian times, *c.*1500 million years ago; (b) Ordovician times, *c.*450 million years ago; (c) Cretaceous times, *c.*100 million years ago; and (d) Quaternary (present-day) times.

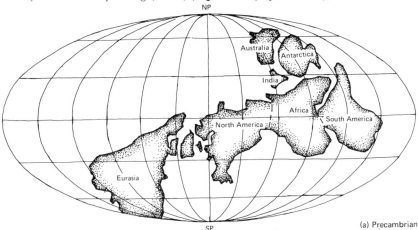

(a) Precambrian

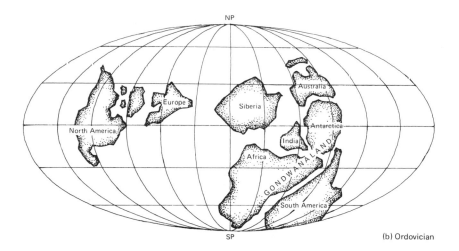

(b) Ordovician

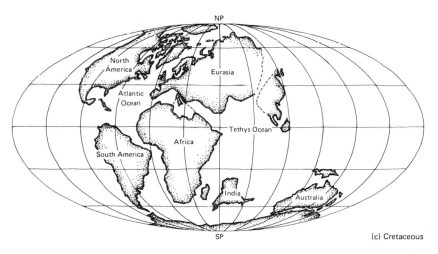

(c) Cretaceous

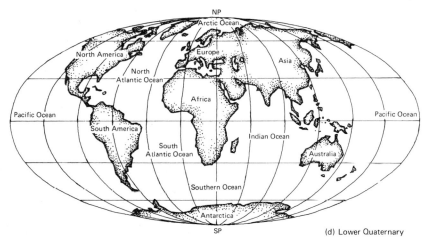

(d) Lower Quaternary

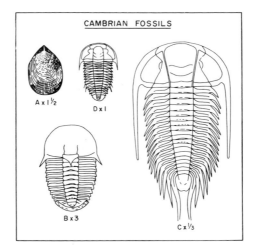

(*Above*) Fossils found in Cambrian strata: A *Lingulella*, B *Hartshillia*, C *Paradoxides*, D *Olenus*. (*Below*) Fossils found in Carboniferous limestone: A *Dibunophyllum*, B *Lithostrotion*, C *Lonsdaleia*, D *Productus*, E *Davidsonina*, F *Zaphrentis*, G *Dictyoclostus*, H *Caninia*.

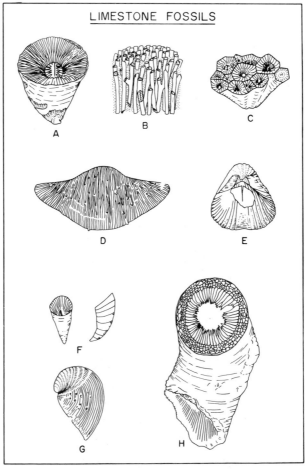

William Smith, between the years 1820 and 1840 the main geological time units were all defined. The leading figures in the establishment of the modern geological timescale were the English geologists Sedgwick and Murchison. They, and others in France and Germany, subdivided the geological column into "systems" such as the Cambrian, Ordovician and Silurian, making use of the by now well established principles of Steno, Hutton, Lyell and Smith. The most important criterion for the recognition of a particular rock system was its fossil content, and as a result the science of palaeontology came of age during the first half of the nineteenth century. Since 1840 there have been only minor changes to the geological timescale.

The main eras and periods were known by their present names in 1840, and the main advances over the past 140 years have been in the recognition of the smaller time and stratigraphic units, referred to as ages and stages. Also there has been a gradual conversion of the relative timescale into a timescale based upon absolute dates.

Landscape and Time

In geomorphology (the science of landscape) the rôle of time has been stressed by many of the subject's most influential authors. Hutton's principle of uniformitarianism found expression in much of the thought of the American W.M. Davis, who dominated the formative period of geomorphology. His greatest contribution to the subject was the "Geographical Cycle", first proposed in 1899 and refined over and again by Davis himself and by his followers for almost half a century afterwards. The basis of this concept was that landscapes evolve through a series of identifiable stages (labelled, for convenience, "youth", "maturity" and "old age") following an initial uplift of the land surface. Under a "normal" climate, such as that of the humid middle latitudes, erosion by running water was thought to be of greatest importance, slowly but surely reducing the land surface to a low, gently undulating peneplain. Climate was thought to be relatively constant; rivers were thought of as more or less permanent features of the environment; and uplift of the land surface was thought to occur only at widely spaced intervals through geological time, with each phase of uplift initiating a new cycle of erosion. Davis himself played down the rôle of climate in the evolution of landscapes, and placed his greatest emphasis on the three factors of structure, process and stage. Inevitably, these three factors did not receive equal treat-

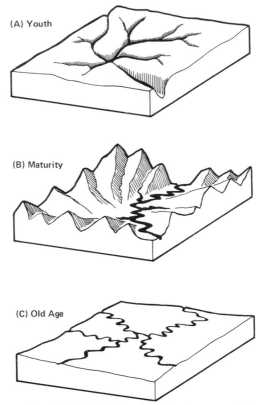

(A) Youth

(B) Maturity

(C) Old Age

Landscapes through time. W.M. Davis' model of landscape evolution called "the
cycle of erosion". No absolute time-scale was suggested, but note the use of terms
used by human beings for recording the stages of their own life cycle.

ment at the hands of the master, and Davis and his followers
became preoccupied with the time element in the equation. In
other words, *stage* became all-important.

In spite of the huge deficiencies of the cycle concept, it did at
least provide a coherent way of looking at landscape, and many of
the early twentieth-century advances in geomorphology were
linked with field studies undertaken within the framework of
Davis' model. Nowadays, however, a cool appraisal of Davis' work
shows that there was too much emphasis on the relative (in other
words, vague) dating of landscape features and on the pigeon-
holing of particular landscape types. Davis' preoccupation with
time caused studies of process to be neglected for several decades,
with the result that many unwise assumptions were made about
the operation of individual processes and about the range of processes responsible for specific landforms and landscapes. There
was also the naïve assumption that the humid middle-latitude

climate was somehow "normal" and that all other climates, such as those of the polar and equatorial latitudes, were somehow abnormal. As late as the 1940s some of Davis' followers were still referring to high-latitude climates as aberrations or abnormalities, and glaciation was looked upon as a "climatic accident". In some ways Davis' thinking was very strongly influenced by Hutton and Lyell, who had emphasized the slow and unspectacular operation of processes through geological time. The cycle concept represented a return to the old idea of cyclic time, with the environment passing through a series of stages and returning to a *status quo*, thereby maintaining in the long term the global balance so beloved of theoreticians.

In the decades following the death of W.M. Davis one school of geomorphology in particular continued to be preoccupied with time. This was the school of "denudation chronology", particularly popular in the United Kingdom during the 1950s. Many students of landscape at this time were concerned above all else with the recognition and dating of old erosion surfaces and with the reconstruction of chronologies of landscape change. Often it was assumed that each erosion surface represented the culmination of a particular cycle of erosion, initiated by uplift and completed through the operation of fluvial processes on the uplifted land surface. Denudation chronology was popularized in the United Kingdom largely through the strength of character of one man— Professor S.W. Wooldridge of London University.

After 1960, studies of erosion surfaces became less common, and, in the reaction which followed, the detailed examination and explanation of processes became much more important in the study of landscape. In the United States and United Kingdom the 1960s and 1970s were the decades of process studies, measurements and quantification. As a reaction against the ideas of W.M. Davis, the theory of long-term cyclic change was replaced by theories of equilibrium and steady state. Some workers became so preoccupied with small-scale process studies and with the idea of "dynamic equilibrium" that they forgot about time and began to interpret landscapes almost entirely in terms of present-day processes continuously or intermittently operating and maintaining an overall state of balance. Some strenuously denied that landscapes consist largely of "fossil" landforms, created in the past and indicative for the most part of past environmental conditions; instead they interpreted every facet of the landscape as a modern feature.

Geomorphology in continental Europe was never very strongly

influenced by W.M. Davis, and a much more important geomorphological theory was the "Treppen concept" of Walther Penck (published 1924). This theory concerned the evolution of landscapes through intermittent tectonic uplift, followed by the formation of certain characteristic hillslope angles and then by the "parallel retreat" of all stable slopes. The end products of parallel slope retreat were thought to be wide terraces, pediments or pediplains with steep-sided erosional remnants standing above them. Classic examples were the "inselberg" landscapes of many hot desert and semi-arid regions.

Penck's ideas were difficult to understand, and they were not widely adopted in the English-speaking world even after his book was translated into English in 1953. Much more popular and influential was the work of Lester King, who also proposed that scarp retreat, resulting in the formation of flat erosion surfaces or pediments, is the standard mode of landscape evolution. He looked upon the semi-arid environment as the "normal" environment, in which the processes of pedimentation operated most fluently. In the theories of Penck and King, time was not of such fundamental importance as it had been to W. M. Davis; although they were concerned with landscape evolution and with the recognition of *stages* as indicators of time elapsed, they were much more concerned with the explanation of *processes* of slope retreat. Like Davis, however,

A typical inselberg landscape with the pediment (flat surface) and steep scarp face or escarpment above. Both Walther Penck and Lester King proposed that scarp retreat was the "normal" mechanism by which landscapes evolve through time.

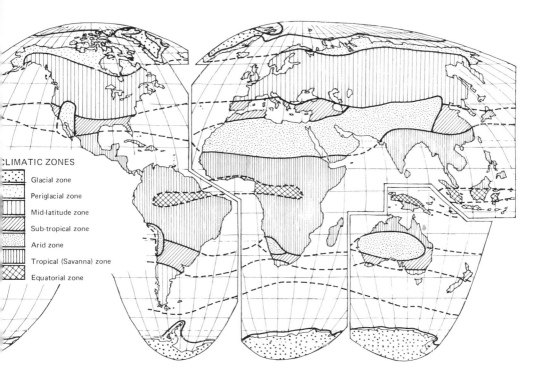

The main climatic zones of the world. Some modern studies of landscape evolution stress that climatic environment is more important than the mere elapse of time in the fashioning of landscape.

they were guilty of making some very naïve assumptions about the stability of climate and about the geomorphological similarity of different environments.

A much more sensitive approach to the environment was made by some French and German scientists. Julius Büdel, in 1948 and 1963, and Professors Tricart and Cailleux in a number of books and papers published in the 1960s, established the main principles of "climatic geomorphology". The essence of this approach to geomorphology is that each climatic zone of the world gives rise to its own peculiar landforms and landscapes. Environment is thus seen as more important than time, at least in pure climatic geomorphology research. In recent years, however, there has been an increasing recognition that climates and environments are no more stable than sea-levels, landscapes, land altitudes or continental arrangements, and most modern studies within the broad field of climatic geomorphology are concerned with the effects of *changing* climate rather than with the assumption of climatic stability.

253

A Typical Subdivision of World Climatic Zones and Landscape Types—the Basis of "Climatic Geomorphology"

Main Climatic Zones	Subdivisions	Main Landscape Types	Characteristic Vegetation
Cold zone	Glacial zone	Glaciers and glaciated landscapes	Antarctic/Arctic desert
	Periglacial zone	Permafrost landscapes and regions of peri-glacial features	Tundra and taiga
Mid-latitude zone	Continental zone	Traces of Pleistocene icesheets; many periglacial features	Boreal forest and cold steppes
	Maritime zone	Many Pleistocene glacial features; fluvial landscapes	Deciduous forest
	Mediterranean zone	Fluvial features and many traces of aridity	Scrub forest
Tropical dry zone	Semi-arid zone (savanna)	Savannas, basins, desert uplands with river canyons, pediments and inselbergs	Warm steppes and savanna woodland
	Arid zone	Sand seas and rocky deserts. Salt lakes	Hot desert
Humid tropical zone	Equatorial or tropical forest zone (selva)	Forested fluvial landscapes, inselbergs and deep bedrock rotting	Equatorial evergreen forest

A Modern View of Time

Both modern geology and modern geomorphology incorporate careful studies of time. The establishment of both relative and absolute timescales is a central (but not all-important) theme, and both research workers and teachers today realize the importance of a balanced approach to studies of the changing surface of the Earth. There is now a recognition that processes do not operate steadily at any scale, but oscillate or fluctuate quite violently in their intensity and effectiveness. At the same time, it is now realized that the climatic environment is always changing, some-times rapidly and at other times so slowly that change is difficult to appreciate or measure.

A recent trend is the recognition of the enormous geomorph-ological work achieved by intermittent events—floods, tidal waves, avalanches, glacier surges, volcanic eruptions and so on. Some might say that this is a return to the days of catastrophism, and

A modern geological catastrophe: this huge debris slide followed the collapse of a mountainside above the Sherman Glacier, Alaska, during the 1964 earthquake there.

indeed some workers (notably the idiosyncratic Immanuel Velikovsky in *Earth in Upheaval* and other books) have continued to preach a crude form of catastrophism right up to the present day. "Neo-catastrophism" is rather more sophisticated, being concerned with time as a central theme and with such problems as the magnitude and frequency of geomorphological events. A modern view of geology and geomorphology would include a recognition of the importance of continuously operating processes, but it would also accept the rôle of intermittent disruptions producing large-scale changes in the nature of sediments or in the appearance of the landscape.

Another important time-related concept is the idea of environmental oscillations of a cyclic nature. Certain sediments appear to be deposited rhythmically or during repeated environmental pulsations: on a short timescale, there are the winter freezes, spring floods and summer droughts which affect the flow (and sediment-

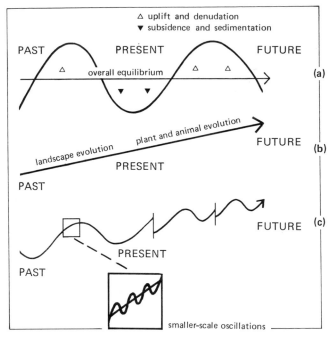

A diagrammatic representation of the three main views of geological time: (a) "cyclic" time, supported by the theory of Uniformitarianism; (b) linear time, supported by, for example, the theory of evolution; and (c) linear time modified by oscillations and disruptions.

transporting capacity) of many of the world's rivers; on a long timescale there are the world's ice ages, recurring more or less regularly through geological time. The recognition of different wavelengths of climatic change is extremely difficult, but here geologists and geomorphologists have a great deal to contribute, and they can cooperate with meteorologists and climatologists in the still youthful field of long-range weather forecasting.

The foregoing paragraphs have outlined some of the attitudes which have appeared in the Earth sciences towards the measurement of time. Chronology is still a central theme, and there is now a preoccupation with the establishment of *absolute* timescales. The relative timescales of the past have established the names of systems, formations and various other units, and those who specialize in time studies are now preoccupied with the fixing of absolute dates to the geological column. A number of techniques have enabled changes of the order of decades to be identified for the past 10,000 years or so, and this level of accuracy is now being extended further and further back in time. And, having established accurate timescales for particular periods, the mathematicians are already enjoying their analyses of "magnitudes" and "frequencies", employing "harmonic analysis" and spending much time communicating with their computers! Our expertise in the measurement of time past is still accelerating at an almost unbelievable pace.

The Age of The Earth

The search for a reliable absolute timescale has been accompanied by many attempts to calculate the age of the Earth. Clearly, the dating of the base of the geological column is critically important, and in the last century the question "How long?" became a central part of the discussions which followed the publications of Hutton, Lyell and Darwin. How long have the unspectacular processes of surface sculpture, such as erosion, had to fashion the face of the Earth? How long have plant and animal species had to evolve through natural selection? How long has Man (who pompously insists on considering himself a "special" creature) inhabited planet Earth?

Early attempts at dating the creation of the planet were bound within the confines of Biblical theology. In 1658 Archbishop James Ussher, Primate of Ireland, set the year of the Earth's creation as 4004BC. This date, calculated from Biblical chronology, was widely—though far from unanimously—accepted by theologians and laymen alike. After all, they had no better method of calculating the span of geological time, and most people were not unduly worried about the problem anyway.

Then, at the beginning of the eighteenth century, the ideas of James Hutton began to make an impact, and scientists became increasingly worried by the brief span of geological (or theological) time as they searched for mathematical order in a rational Universe. Hutton and Lyell talked of almost limitless time, time to come and "the boundless mass of time already elapsed". Various crude attempts were made to calculate the age of the Earth using rates of sediment deposition and rates of salt deposition in the oceans. One estimate based on the salinity of the oceans gave a figure of 90 million years, and estimates in the late 1800s based on deposition rates generally varied between 20 million years and 100 million years. One or two exceptional estimates gave ages of over 1,500 million years, but these were thought by the scientific establishment to be wildly inaccurate and not worth considering. After all, Charles Lyell, as one of the most influential of all nineteenth-century geologists, set the span of geological time as 240 million years, and most people thought of this estimate as on the high side.

William Thomson, Baron Kelvin, (1824–1907) the brilliant physicist whose estimate of the age of the Earth was 70 million years, an estimate which did little to enable people to understand the sheer immensity of geological time.

One of the major constraints upon the dating of the Earth was the involvement of Lord Kelvin in the debate. Kelvin was a highly respected physicist who devoted much time to calculating the age of the Sun and the rate of cooling of the Earth. His work was not, so far as could be seen, dependent upon dubious assumptions, and because his calculations were mathematically exact and apparently infallible they were widely accepted. His maximum estimate for the age of the Earth was 70 million years, and he concluded in 1897 that the Earth had probably been habitable for between 20 and 40 million years. This seemed to be the last word in the discussion, and the followers of Hutton, Lyell and Darwin were dismayed to say the least. How could the Earth's surface have been fashioned in such a short time if the principle of uniformitarianism had any meaning? How could natural selection have operated over a period as short as 40 million years? Perhaps the catastrophists were right after all . . .

The uncomfortable compression of the geological column did not last long. In 1896 Henri Becquerel discovered radioactivity, and by 1902 it was known that radioactive processes maintain the energy output of the Sun and the temperature of the interior of the Earth. No longer was it necessary to look upon the Sun and the planets as heavenly bodies cooling down and running out of energy. Suddenly, an immensely ancient Sun and a slightly more youthful Earth were distinct possibilities, and radioactivity seemed to hold the key to absolute dating. By 1905 it was known that radioactive elements decay at set rates, often changing into other elements in the process. The discovery that lead was a stable end-product of the decay of uranium held the key to radioactive dating, and B.B. Boltwood showed in 1907 that in rocks of different ages the ratio of lead to uranium differed by a predictable amount. Boltwood calculated the following radiometric ages for rocks of known "relative" age as follows:

Carboniferous rocks: 340 million years old;
Silurian or Ordovician rocks: 430 million years old;
Precambrian rocks (Sweden): 1270 million years old;
Precambrian rocks (Ceylon): 1640 million years old.

These were the dates which marked the end of the era of "naïve dating". From now on there could be no doubt about the immensity of geological time, and Earth scientists could begin to fit their strata and their fossils, their ice ages and their periods of marine incursion into an absolute scale of geological time.

Arthur Holmes (1890–1965), the British geologist whose work on radiometric dating showed that the Earth was not tens but thousands of million years old.

Since 1907 a geological timescale based on radiometic age determination has been gradually refined. Among the first to construct
such a timescale was the British geologist, Arthur Holmes, a
pioneer in modern dating techniques. Between 1911 and 1947 he
published a number of improvements to the geological timescale,
but he had inadequate resources for the dating of large numbers of
samples. Up to 1960 there were several substantial adjustments
in the dating of the boundaries between the different geological
systems, but in 1964 a more or less definitive version of the timescale was published by the Geological Society of London. Since
that time, in spite of a large increase in the number of radiometric
dates published for various parts of the geological column, adjustments in the dating of critical boundaries have been only minor.

The geological timescale shown in the diagram is the latest
version available, demonstrating the immense antiquity of many of
the rock formations of the Earth. The oldest known terrestrial
rocks are about 3,800 million years old, but iron meteorites have

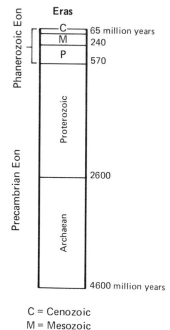

C = Cenozoic
M = Mesozoic
P = Palaeozoic

The geological timescale subdivided to scale. The Precambrian, about which we
know very little, occupies the large majority of the column; the last 570 million
years, which is to us an unrealizably vast timespan, is squashed up near the top of
the column. Those who wish to plot the time since the emergence of manlike
beings are challenged to draw a line 1/1150th (at most) of this diagram's height.

been dated as about 4,600 million years old, and so have the oldest moon rocks. On this basis it is widely assumed nowadays that the Earth and the other planets were formed about 4,600 million years ago, perhaps slightly earlier. This date marks the beginning of *geological time*; for events which occurred prior to 4,600 million years ago we should strictly refer to *cosmic time*, which we met in Chapter 5, for which a suitable scale still has to be evolved.

Lest we should begin to feel that our view of geological time is now enlightened and strictly rational, we should remember that we know remarkably little about 85% of the time which has elapsed since the birth of planet Earth. Our view of time is still a wildly distorted one, and we can be sure that there are still any number of surprises in store for Earth scientists as they investigate the events represented in the lower parts of the geological column. This statement from Don L. Eicher may help us to see things in perspective: "Compress the entire 4.6 billion years of geologic time into a single year. On that scale, the oldest rocks we know date from about mid-March. Living things first appeared in the sea in May. Land plants and animals emerged in late November and the widespread swamps which formed the Pennsylvanian [later Carboniferous] coal deposits flourished for about four days in early December. Dinosaurs became dominant in mid-December, but disappeared on 26th, at about the time the Rocky Mountains were first uplifted. Manlike creatures appeared sometime during the evening of December 31st . . ." (*Geologic Time*, Prentice Hall Inc., 1968, p.19.)

Geological Dating Methods

There are now many dating techniques available to the Earth scientist for the establishment of absolute rock ages. The most important of them are the radioactive dating methods developed from the pioneering work of Holmes and others. However, most of the techniques employed so far measure the decay of radioactive materials in rock, assuming that the parent atoms were created at the same time as the rock itself. This assumption is reasonable in the case of most igneous rocks and some metamorphic rocks; but, since sedimentary rocks are usually made of materials derived from the breakdown of older rocks, most of the radioactive materials which they contain are older than the sediments themselves. On the other hand, there are a few useful isotopes which are contained within living and dead organisms such as corals and marine animals, and sediments which contain organic remains in abundance can sometimes be dated radiometrically.

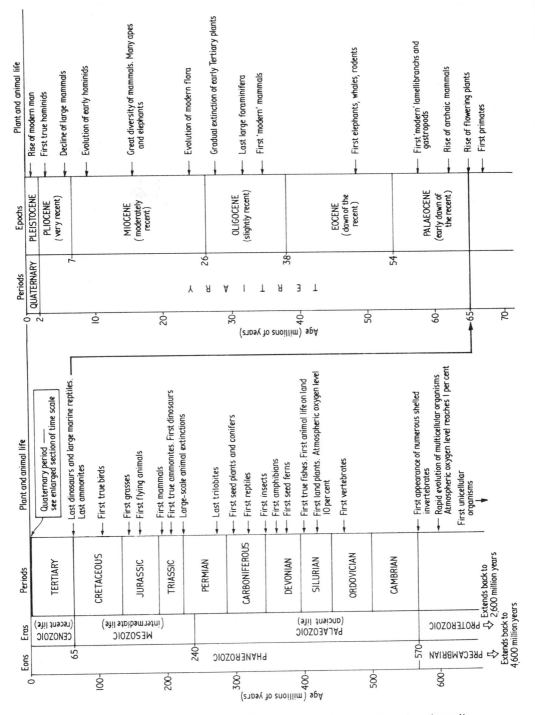

Geological column showing the main eons, eras and periods, based on the radio-metric timescale. The right-hand column shows the most important stages in the evolution of plant and animal life.

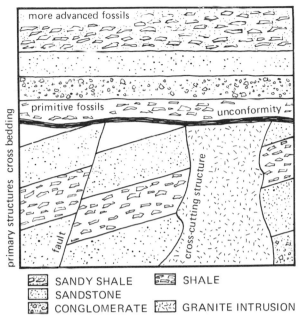

more advanced fossils

primitive fossils

unconformity

primary structures | cross bedding

fault

cross-cutting structure

SANDY SHALE	SHALE
SANDSTONE	
CONGLOMERATE	GRANITE INTRUSION

A cross-section through a hypothetical sequence of strata. Here we can see a number of stratigraphic relations and other features of a rock sequence which might help in the relative dating of the individual rock features. Note the unconformity across the centre of the diagram: this shows that there has been some erosion of the lower sediments before the sediments above were laid down.

Inference and Correlation

Today there is a sound framework of absolute rock ages for the geological column, especially for that part of it referred to as the Phanerozoic (younger than about 570 million years). And yet it still remains true that most rock formations and sediment layers are dated by inference and correlation, using relative dating techniques based upon the old-established principles of stratigraphy and palaeontology. A field scientist confronted by a sequence of rocks or sediments can interpret the age of a particular formation by noting some of the following characteristics as appropriate:

Lithology: the internal characteristics of the formation, such as texture, colour and internal composition. This indicates whether the rock is of igneous, metamorphic or sedimentary origin.
Primary structures: these include layering, cross-bedding, graded bedding, ripplemarks and other features concerning the arrangement of minerals and rock fragments.
Stratigraphic relations: these refer to the sequence of strata or the relations of the rock formation in question with overlying or underlying formations.

Unconformities: breaks in a sequence of strata which indicate that not all of the original layers are still present. Often a period of erosion removes part of a rock sequence, only for the erosion surface itself to be buried by more rock layers as sedimentation continues. Sometimes tilted or deformed rocks are separated by such an erosion surface unconformity from overlying, younger rocks which are less deformed.

Cross-cutting structures: these are either igneous intrusions (such as dykes and batholiths which cut indiscriminately across previously existing strata), or else faults, which disrupt the bedding or layering of rock. Always these structures are younger than the rocks which are affected.

Fossils: if there is any evidence of prehistoric life in a rock sequence, the principles of palaeontology can be used for the relative dating of fossiliferous rock layers. Dating is done on the basis of the types of life forms represented, the abundance of certain species, etc.

Using the above principles, a skilled geologist can often estimate the age of a particular rock formation to within about 20 million years. Normally he is able to date the rock to a specific geological period, or even to a particular epoch. In some cases, where a rock formation has a distinctive character (for example a striking red colour in a suite of rocks which are for the most part coloured grey) or if it has a contained fauna of distinctive "index fossils", dating may be possible to within 5 million years. If, nearby, a particular rock formation is overlain by a lava flow which can be dated by radiometric means, a very accurate date may be assigned to our rock formation even though it itself is not overlain by the lava flow. In this case the geologist must, of course, be certain of the *lateral continuity* of the formation, for there are many instances of rocks having strong physical similarities but widely differing ages.

In many areas of complicated geology and structure it is difficult to assign "time labels" to suites of rocks simply by using the principles described above. In such cases a number of other methods are available for the correlation and dating of rocks.

Palaeoclimatic methods are often employed. If a geologist is able to recognize a sequence of climatic or environmental changes in the rock formations being studied, he may be able to "match up" this sequence of changes with nearby areas where the stratigraphic succession is better known. Thus, if he finds that the rock facies and fossils indicate clear warm-water conditions followed by a shallowing of the sea and by a sequence of environments ranging from

sandy beach to muddy shallow water and river delta, he may know
by comparison with other areas that he is dealing with the
boundary between the Carboniferous Limestone series and the
overlying Millstone Grit series.

Tectonic methods of relative dating are also useful. There have
been a number of major mountain-building episodes at intervals
through geological time, and the deformation of rock strata can
often be referred to one or other of these episodes. Sometimes the
trend of folding in the rocks gives the clue to the age of the episode
involved. Thus rocks in west Wales which have fold axes aligned
north-east to south-west can be dated as Silurian or older, as the
direction of folding is typical of the Caledonian orogeny which has
been dated to about 395 million years ago. On the other hand,
rocks which have fold axes trending north-west to south-east could
be as young as 280 million years old, since they were affected by
the later Armorican or Hercynian phase of mountain-building.

Geophysical techniques are now available for identifying rocks
which are not visible at the ground surface. Some instruments
enable the operator to recognise strata on the basis of their electrical conductivity; others track the passage of seismic waves through
rock following the carefully controlled explosions of small
dynamite charges. Deflections of the seismic waves recorded on
"seismograms" can be correlated with particular rock units. Using
these and other techniques, skilled operators can identify the
imprint or character of a particular rock of known age without
having any physical access to the rock itself, so that geophysical
techniques can in themselves become useful dating tools.

A particularly useful method of dating is based upon the fact
that the Earth's magnetism is recorded in rocks. Under normal
conditions, the magnetic field is directed towards the north; the
inclination of magnetism to the horizontal is high in polar areas,
parallel with the surface at the equator, and intermediate at inter-
vening latitudes. Thus, from the angle of dip of the magnetism in a
rock sample, the geophysicist can obtain information about the
latitude at which the rock was formed. (Because of continental drift
and other processes this "palaeolatitude" may be quite different
from the latitude at which the sample was collected.)

The magnetic characteristics of a single sample cannot be used
for dating, but a great deal of suggestive evidence can be gained
from the magnetic characteristics of a rock sequence from a site.
This is because the Earth's magnetic field is intermittently
reversed, so that at times a compass needle would point south

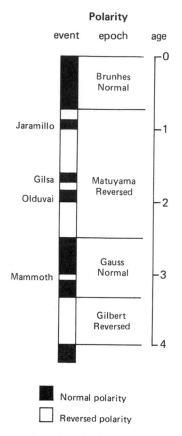

Polarity

event	epoch	age

Brunhes
Normal

Jaramillo

Gilsa
Olduvai

Matuyama
Reversed

Mammoth

Gauss
Normal

Gilbert
Reversed

■ Normal polarity

□ Reversed polarity

Work done in the last couple of decades has shown that the Earth's magnetic field periodically reverses in "direction" (such a change would mean that the "north" end of your compass needle would point south). This is a very simplified geomagnetic timescale for the past 4.5 million years, showing main polarity epochs and short-lived "events".

rather than north; such reversals are tens or hundreds of thousands of years apart.

There is now a comprehensive record of rock magnetization for the last 4½ million years, and if a pattern of magnetic anomalies can be obtained for a particular rock sequence this can be matched up with the pattern of anomalies on the geomagnetic timescale. This timescale is now well established through radiometric dating of lavas; thus a date for the "matching" segment of the timescale can normally be obtained for the rocks being investigated.

Radiometric Dating
As mentioned above (p. 260), radiometric dating is now the technique of greatest importance for establishing absolute rock ages. Of

the dozens of radioactive nuclides which occur in nature, just four have provided almost all the radiometric ages for ancient rocks. These are uranium-235, uranium-238, rubidium-87 and potassium-40. Other radioactive nuclides are not widely used because they are too rare, or because they decay at too slow a rate to be of use for even the most ancient rocks, or because they decay too rapidly.

The critical characteristic of an isotope used in rock dating is its *half-life*. Each isotope has a unique half-life and a unique and complicated process of "daughter" nuclide (that is, the nuclide created by the decay of the original nuclide) production. Normally the basis of the dating method is the measurement of the ratio of parent nuclides to daughter nuclides present in the rock; it is

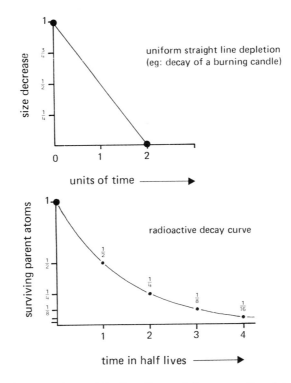

A comparison between the straight line of normal decay over time (*above*) and the curve of radioactive decay (*below*). In the upper diagram, imagine that time is being measured using the height of a burning candle: a centimetre's decrease indicates (approximately) a measurable amount of time which is the same no matter what the height of the candle. In the lower diagram, the idea of a half-life is shown: whatever the *initial* amount of the substance present, in a certain unit of time half of it will have decayed.

The Chief Methods of Radiometric Age Determination

Parent Nuclide	Half-life (years)	Daughter Nuclide	Minerals and Rocks Commonly Dated
Uranium-238	4,510 million	Lead-206	Zircon, Uraninite, Pitchblende
Uranium-235	713 million	Lead-207	Zircon, Uraninite, Pitchblende
Potassium-40		Argon-40	Muscovite, Biotite, Hornblende, Glauconite, Sanidine, whole volcanic rock
Rubidium-87	47,000 million	Strontium-87	Muscovite, Biotite, Lepidolite, Microline, Glauconite, whole metamorphic rock

always assumed that when the rock was formed there were only parent nuclides present and that the process of decay and daughter nuclide production commenced immediately.

Uranium-lead dating is now widely employed in the determination of rock ages. All uranium which occurs naturally contains uranium-238 and uranium-235, and these two separate radioactive nuclides provide a cross-check in determining rock ages. Uranium-lead dating was first applied to uranium minerals like pitchblende and uraninite, but these minerals are so rare that the technique was restricted in its use. When delicate methods of measuring minute quantities of uranium and lead were evolved, it became possible to use the widespread mineral zircon, and this greatly broadened the potential of the uranium-lead method to include igneous rocks from many different regions. As a result of radioactive decay, uranium-bearing minerals continuously accumulate lead, and the accurate measurement of the amounts of uranium and lead present in the material is the basis of the dating method. The relative proportions of the individual isotopes are measured, and rock ages can be calculated from the ratios of uranium-238 to lead-206, uranium-235 to lead-207, or lead-206 to lead-207. As an additional check, the ratio of thorium-232 to lead-208 can be calculated, and if the ages calculated for a single sample using each of the ratios agree approximately, they are said to be *concordant* and probably correct.

Potassium-argon dating is useful for the dating of all types of rocks containing potassium-bearing minerals, such as biotite, muscovite, hornblende and glauconite. In addition, whole rocks

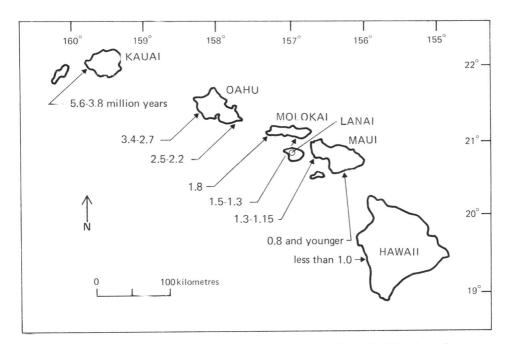

The potassium-argon method has been used to date these Hawaiian islands, and shows that the oldest island is in the northwest, the other islands becoming progressively younger toward the southeast.

may be dated. The naturally occurring radioactive potassium isotope is potassium-40, which decays to form calcium-40 and argon-40. The calculation of the ratio between potassium-40 and argon-40 provides dates for ancient rocks. For young volcanic rocks, minute quantities of argon-40 can be measured by examining the whole rock. Reliable dates as young as 100,000 years have been obtained, and under ideal conditions it should be possible to date fine-grained lavas as young as 40,000 years.

The rubidium-strontium method of dating is based upon the measurement of the ratio of the radioactive isotope rubidium-87 to its daughter product strontium-87. Micas and potassium feldspars are the most suitable minerals for rubidium-strontium age determinations, and the results can commonly be compared to potassium-argon dates for the same samples. Most of the rocks dated are ancient igneous rocks, but this method appears to be particularly useful for the dating of metamorphic rocks (those rocks, originally either sedimentary or igneous, which have been "metamorphosed" by heat and/or pressure during submersion within the Earth's crust). Sedimentary rocks containing glauconite can also be dated.

There are two main methods of dating marine sediments from the recent geological past. These are the thorium-230 method (providing dates up to several hundred thousand years old) and the thorium-230/protactinium-231 method (useful for ages up to 150,000 years).

Thorium-230 is a decay product of uranium-238, and in the sea it is quickly precipitated and incorporated into the sea-floor sediments. It decays with a half-life of 75,000 years, and ages are calculated by measuring the concentrations of thorium-230 at particular depths in deep-sea cores and comparing these with the surface concentration.

Protactinium-231 is a decay product of uranium-235 which also precipitates quickly in the sea. It has a markedly different decay rate from thorium-230, however, so that a thorium-protactinium ratio calculated for a particular sample from a deep-sea core and compared with the ratio at the sediment surface can be used as the basis for dating. These two dating techniques have been of great importance during the last two decades as a result of the intensive studies of deep-sea cores for clues to Quaternary climatic change.

Studies of long sediment cores from all of the deep oceans reveal systematic changes in the assemblages of planktonic (surface) and benthic (sea-floor) foraminifera, and a number of methods have been devised for recognizing changing environments during the deposition of columns of sediment from examination of fossils of these tiny sea animals. For example, the proportions of warm-water and cold-water foraminifera can be used to estimate the water temperature at which a certain layer was deposited. One species of foraminifera, *Globorotalia truncatulanoides*, has a unique characteristic of coiling to the left in cold water and to the right in warm water, so that a plot of the ratios of left- to right-coiling forms gives an indication of the water temperatures when different levels in a deep-sea core were deposited.

Another widely used method of measuring the changing temperatures at which deep-sea sediments were laid down concerns the ratio between the stable isotopes oxygen-16 and oxygen-18. The heavier isotope oxygen-18 is most abundant in calcium-carbonate shells when water temperatures are relatively warm; on the other hand, a relative decrease in oceanic oxygen-18 can also be used to demonstrate the addition of great quantities of glacial meltwater to the oceans, thereby cooling them and indicating a change from glacial to interglacial conditions during an ice age (we live at the moment during an interglacial of an ice age that has

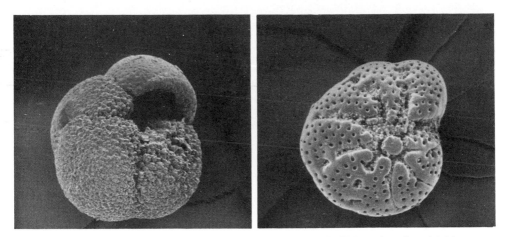

Clues about climate. Two Pleistocene foraminifera: (*left*) *Globoratalia inflata*; (*right*) *Elphidium oceanis*.

been in progress for some millions of years). Conversely, when an ice sheet is building up during a glacial stage, the proportion of oxygen-18 to oxygen-16 increases, since the water-vapour which is converted into snowfall takes up oxygen-16 more easily than oxygen-18.

There are many difficulties and sources of uncertainty in the interpretation of oxygen-isotope ratios, not least of which is the problem of assigning absolute ages to key points on a typical curve of deep-sea temperatures. The study of oxygen-isotope ratios in deep-sea sediments cannot in itself provide information on absolute ages, so other techniques such as uranium-isotope dating and geomagnetic dating have to be employed.

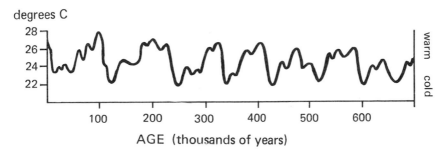

A generalized curve of temperature changes over the past 700,000 years, based upon the analysis of deep-sea sediments.

A modern radiocarbon dating laboratory, showing some of the equipment used for the purification of the gas containing the radioactive isotope carbon-14.

Measuring the Age of Man

Radiocarbon Dating

One of the most widely employed techniques for measuring time past is radiocarbon dating, originally developed by W.F. Libby. It is based upon the measurement of amounts of the radioactive isotope carbon-14 in dead organic materials such as wood, bone, marine shells, peat and organic mud. Carbon-14 is present in living things in all sorts of environments, and the quantity of the isotope is in balance with the quantity in the environment itself. As soon as an organism dies the contained carbon-14 decays with a half-life of 5,570 years. Because the half-life is so short, the measurement of residual carbon-14 is a very delicate matter, and even with the most refined of instruments the limit of radiocarbon dating is about 40,000–50,000 years BC. However, this covers that part of the geological column about which we know most, encompassing the most recent part of the Quaternary ice age, the evolution of modern floras and faunas, the recent evolution of Man himself and the great flowering of his many cultures. Inevitably, therefore, radiocarbon dating has been widely employed in a broad range of sciences, from geology and geomorphology to pedology, archaeology and even medieval history. The radiocarbon age scale is now widely accepted as the basis for most studies of chronology over the past 40,000 years or so, and, although the dating method has its own peculiar sources of error, refinements are being introduced all the time which increase its accuracy and reliability.

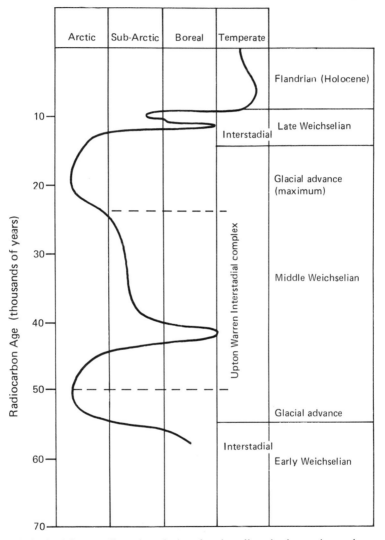

Graph devised from radiocarbon dating showing climatic change in northwest Europe since the last interglacial: time reads upwards, temperature across. The European "Weichselian" of this diagram is equivalent, approximately, to the North American "Wisconsin".

Other Dating Techniques

There are a number of other dating techniques which have been employed for dating at the "archaeological end" of the geological timescale. However, it should be realized that a number of these methods are dependent upon accurate calibration before they can be thought of as providing "absolute" ages for sediments or culture levels. Very often radiocarbon dating or potassium-argon dating is employed in the calibration process.

The main techniques which are independent of organic materials are varve dating and tephrochronology. The former method was pioneered by the Swede Gerard De Geer, and later applied by E. Antevs and others in North America. It involves the counting and correlation of varves or annual layers of fine sediment deposited in water near the margins of wasting ice sheets. In his original study of Swedish varves De Geer measured varve thicknesses at many different sites, plotting these thicknesses on a timescale and visually matching the sequences from adjacent sites. Gradually he built up a varve sequence over a distance of almost 1,000km, and he managed to establish a chronology stretching back over nearly 17,000 years. Since 1960 radiocarbon dating has been used to check De Geer's chronology, and it has been shown to be remarkably accurate.

Tephrochronology is the study of ash layers of known date with a view to establishing a reliable chronology for an area affected by volcanic processes. Much of the pioneer work was done by S. Thorarinsson in Iceland, where there are historical records of eruptions for the last thousand years or so. Particular ash layers (which may be no more than a few millimetres in thickness) can be distinguished on the basis of their mineralogy and colour, and, if correctly identified, they can be used as very accurate "marker horizons" in sequences of sediment.

One of the most important of all the methods used in elucidating the climatic and environmental changes of the last half million years or so is palynology or pollen analysis. The method has been used particularly for the present interglacial, but it has also been used successfully for the separation and dating of layers of sediment from the earlier interglacials of the Pleistocene period. The basic assumption of the method is that the pollen grains preserved in sediments give a reasonable picture of the character of the local vegetation of the time. Pollen grains are remarkably resistant, and since they can survive in sediments such as peat and organic mud for many thousands of years there are no great worries about the fossil pollen assemblage being different from the original assemblage. By identifying and counting the pollen from the layers in deposits, palynologists create pollen diagrams which show the

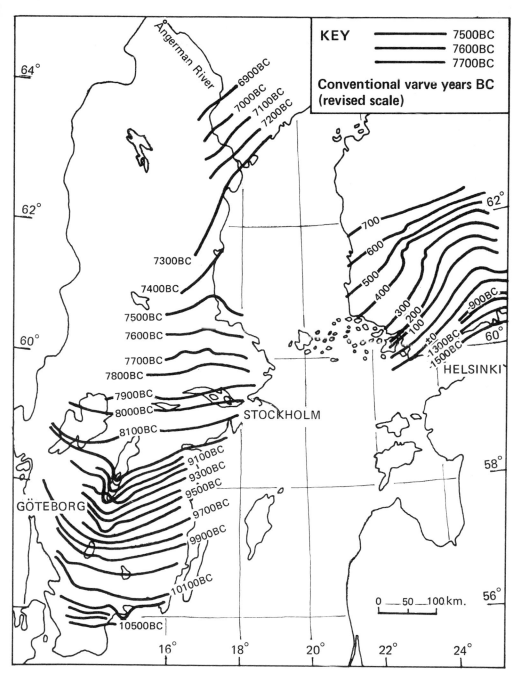

Conventional varve years BC
(revised scale)

Map of the stages of ice-margin retreat across Sweden and Finland at the end of the last glaciation, based on De Geer's counting of varve sequences. De Geer's time calculations were remarkably accurate, and have been largely confirmed by radiocarbon dating. The Finnish "zero year" is 8,200 BC.

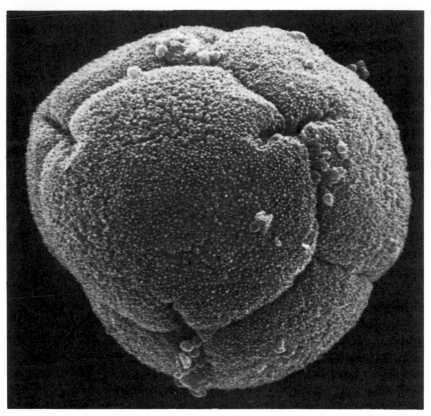

A typical pollen grain found in interglacial sediments—*Empetrum nigrum L* at a magnification of × 4000.

composition of the various floras in the sequence. Having done this, they can recognize the layers of typically warm-climate and typically cold-climate pollen assemblages, and also layers which indicate increasing water salinity, increasing forest cover and so forth. Zones which mark the times of climatic change can also be identified. Now, after many years of study, the "typical" pollen assemblages from the various glacial and interglacial stages of Europe and North America are well known. Calibration has also been achieved through the use of radiocarbon dating and other absolute dating methods, so that newly investigated sediments can often be dated quite reliably simply on the basis of their pollen assemblages.

As indicated earlier in this chapter, other fossils are also widely used for the dating of sediments, even if they are unconsolidated. Recent sediments may contain marine molluscs, animal bones, horns and antlers, fish remains, twigs and leaves, and even insects

Subdivision of the Present Interglacial of the British Isles into Pollen Zones (The age scale in the left refers to radiocarbon ages)

Time based on C^{14} dating	Periods	Pollen zone	Zone characteristics for England and Wales		
1000 AD	Sub-Atlantic	VIII modern	Afforestation		
		VIII	Alnus-Quercus Betula (-Fagus-Carpinus)		
BC 1000 2000 3000	Sub-Boreal	VIIb	(Deforestation) Alnus-Quercus-Tilia Ulmus decline		
4000 5000	Atlantic	VIIa	Alnus-Quercus-Ulmus-Tilia		
6000	Boreal	VI	Pinus-Corylus	(c) Quercus-Ulmus-Tilia (b) Quercus-Ulmus (a) Ulmus-Corylus	
7000		V	Corylus-Betula-Pinus		
8000	Pre-Boreal	IV	Betula-Pinus		

such as beetles. All of these have been used in detailed studies of environmental change, for every plant and animal species has its own preferred habitat; some species have particularly sensitive environmental requirements, indicating air or water temperatures to within a few degrees or demonstrating that certain moisture or salinity requirements were matched precisely at the time the sediment was formed. As with pollen analysis, once a scale of species change has been established for a region and once this scale has been properly calibrated, newly discovered fossils can often be dated quite accurately by correlation.

Dendrochronology was pioneered by the American H.C. Fritts. The method involves the counting and measuring of annual tree rings in regions where there is a wide variation between winter and summer climates in each year. Variations in the thickness of rings are related above all to the changes in climate over periods of years, and "patterns" or sequences of thick or thin rings can be matched up from tree to tree. Using living and dead bristlecone pine trees from the south-western USA, Fritts and his co-workers have been able to extend their timescale back to about 5000 BC. Like the varve chronology, the tree-ring scale is theoretically "absolute" and should be able to stand on its own. However,

A fossil treestump preserved in marine muds below sea-level on the coast of Wales, UK. Fossils such as this can be used as checks for carbon-14 dating, and also provide information about the position of sea-level at the time of growth.

because of the risks of inaccurate correlation from tree to tree, and of counting "false" rings which do not represent the passage of a year, calibration by reference to historical data or by radiocarbon dating is often required.

Lichenometry is a technique involving the measurement of certain parts (the thalli) of certain lichen species which have colonized rock surfaces. The basic idea, enunciated originally by R.E. Beschel, is that once a rock surface is exposed to the atmosphere (following, for example, the uplift of a shoreline or the retreat of a glacier margin) it will be colonized by lichen. In theory, the diameter of the largest lichen thallus in an area is proportional to the length of time over which the surface has been exposed. The usual lichen species used is *Rhizocarpon geographicum*, which has a broad distribution, a long life, and a nearly constant age-size relationship. It also produces thalli which are nearly circular, and this helps greatly in measurement. Although many successful lichenometry studies have been completed in areas that have recently been deglaciated, there are many risks involved in the technique, and radiocarbon calibration is always needed before absolute ages can be assigned with any confidence.

(*opposite*) Bristlecone pine trees from southwest United States have been used for much of the pioneering work in dendrochronology.

Fluorine dating is a useful technique in the study of buried bones. The principle of the method is that underground water, as it percolates through the rocks and soil, carries minute quantities of fluorine, which gradually replaces the calcium in buried bones. The change is irreversible, and by comparing the ratio of fluorine phosphate to calcium phosphate in the bones an approximate age can be determined. Old and young bones at a particular site can easily be differentiated by using this method.

A relatively new technique of dating is amino acid dating, now being used on marine shells, bones, foraminifera and even plants. The principle behind the method is that with time there is a change in the optical configuration of amino acids held in the protein

structure of the fossil. However, since the chemical reaction involved is sensitive to temperature the precise criteria for calculating absolute ages will vary from site to site. Nevertheless, the technique holds great promise for the dating of organic materials which are too old for radiocarbon dating.

In addition to the above techniques there are very many others which have been used over the years in attempts to measure time past. Some of these techniques are of strictly local application; some have proved so unreliable that they have been abandoned; some provide information only on relative ages, and have to be supplemented with other dating methods; and some are new techniques which will not fulfil their promise until new and sophisticated measuring devices are developed. But all of them contribute to the search for an absolute and foolproof timescale, and seen in this context *every* available method needs to be tried and tested.

Archaeological Tools
Many of the dating techniques employed in modern archaeology are closely similar to those described in the foregoing pages. For

A "cromlech" or megalithic chambered tomb. This is the core of the tomb, now exposed following the removal of its covering earth mound. Features of this type are valuable for the dating and correlation of cultures from widely dispersed areas.

Radiocarbon Age Scale for the Main Archaeological Divisions of the Past 100,000 years or so in the British Isles

Stage	C^{14} age B.P.	Culture, industry or stage	Division of Stone Age
Flandrian (interglacial)	3,000–1,000	Iron Age	
	2,000–2,600	Late Bronze Age	
		Early Iron Age	
	2,700–3,900	Middle Bronze Age	
	3,000–3,800	Early Bronze Age	
	3,500–4,000	Beaker ⎱	Neolithic
	3,500–5,300	Windmill Hill ⎰	
	6,000–7,000	Sauveterrian ⎱	Mesolithic
	9,550	Maglemosian ⎰	
Devensian (glacial)	6,000–10,000	Creswellian ⎱	Upper Palaeolithic
	20,000–30,000	Proto-Solutrean ⎰	
		Mousterian	
Ipswichian (interglacial)		Lavalloisian	Middle Palaeolithic

example, the principles of stratigraphy are basic to archaeology and to the relative dating of successive cultural horizons. Unconformities can be interpreted and dated in cultural successions in exactly the same way as in rock sequences. Type fossils (which may be organic remains or artifacts) can be used for the dating of horizons or levels in deposits which are associated with various cultures. The typology of artifacts can be used as an aid to dating, just as plant or animal morphology is used for relative dating in the science of palaeontology. And, just as divergent and convergent evolution occurs in plants and animals, so development and "seriation" can be traced in human artifacts. Here again, assumptions can be made about the relative or absolute dating of features on the basis of their forms and supposed functions. Age information can be gained also from the study of artifact distribution, especially if it can be proved that an artifact originated in a particular locality and then spread outwards by the processes of cultural diffusion. In archaeology cross-dating is used to demonstrate the contemporaneity of cultural groups if there is no independent chronology to be referred to. The principles used are exactly the same as those employed in geology for the correlation of strata or layers of sediment which are not demonstrably continuous.

The most important absolute dating technique in archaeology is radiocarbon dating, and many dates have been obtained for the cultures which have flourished within the past 40,000 years or so—in other words, since the onset of the last major glaciation of the northern hemisphere. For periods beyond the range of radiocarbon dating other techniques have to be used. For example, in

A reconstruction by Maurice Wilson of *Australopithecus*.

elucidating the story of the evolution of Man in Africa the following techniques have been important: potassium-argon dating, geomagnetic dating and uranium series dating. In some instances other radiometric methods have been used, including the thorium-230 method and the thorium-230/protactinium-231 method. Additional relative dating techniques involving the analysis of floras and faunas, the reconstruction of palaeoclimates, the correlation of strata and the interpretation of erosional phases have also been essential in the construction of the prehistoric timescale. As a result, Man's earliest ancestors (the various species of *Australopithecus*) are now reliably dated to at least 5 million years ago. Hominids were not only present during the Pleistocene period, but in Africa they existed for at least the whole of the Pliocene period as well. Man's antiquity is established beyond all doubt, and he was certainly in residence on planet Earth well before Adam and Eve—according to Archbishop Ussher!—enjoyed the fruits of the Garden of Eden.

Archaeomagnetism is one of the specifically archaeological techniques now employed for the dating of past cultures. It is based on the fact that, when a magnetic oxide of iron is cooled after heating, its magnetism is determined by the magnetic field in which it lies. Such oxides occur naturally in nearly all clays, and when a clay kiln or hearth is cooled and abandoned it will retain magnetic properties which provide precise information about the site: these are the magnetic declination, dip and intensity. These all vary with time, and in some areas curves showing magnetic variations over the past 2,000 years or so are now available. By reference to this curve, newly discovered hearths can be dated to within 50 years.

Thermoluminescence is a technique for measuring the light emitted from mineral crystals following radiation and heating. Normally the analysis of thermoluminescence is undertaken on ancient pottery, but much development is needed before this method will be perfected.

Obsidian dating is employed for the measurement of time elapsed since the exposure of a fresh surface of obsidian to the atmosphere. The basis of the technique is that change occurs at a very slow constant rate as water is taken into the material's structure. The rate varies with temperature but not with the quantity of water available, so dating can be undertaken by comparison with other artifacts from the same climatic region. The thickness of the hydration layer on an artifact is measured optically in a thin section, and by reference to a known local scale it can then be converted readily into age in years.

Where bones need to be dated and where fluorine dating is impracticable, collagen dating may be used. Animal bone consists basically of calcium phosphate associated with two organic materials, bone protein or collagen, and fat. After the death of the animal the fats break down rapidly and disappear. The collagen survives much longer, although in decreasing amounts, and it can be measured by analysis of the nitrogen present. Since the method can be used only for fixing the relative ages of bones, it is usually used in association with radiocarbon dating.

Coda

It is worth concluding this chapter with a few words of caution. We fondly assume nowadays that the old methods of relative dating have been supplemented if not supplanted by the new methods of absolute dating using the gadgetry and the expertise of today's technological society. We assume all too often that the radiometric time scale is *correct*. And all too often we refer to radiometric ages as absolute ages. At the end of his book *Geologic Time* Don Eicher says this: ". . . suspicions have persisted that, because of some unanticipated source of systematic error, the whole radiometric calendar from bottom to top might be drastically wrong . . ." (p. 139).

Already it has been shown that some of the early assumptions of radiometric dating were incorrect, and today a number of variables are fed into the calculations of radiometric ages which were not even thought of a decade ago. It has been shown that the specifications of the equipment used in radiometric dating can significantly affect the results obtained. In radiocarbon dating one of the basic assumptions has always been that the rate of production of carbon-14 in the atmosphere has been uniform throughout time past. This assumption is now known to be incorrect, and radiocarbon dates of 4,500 years ago may be in error by about 700 years. The error increases progressively before that date, and although tables are now available for the correction of dates there are still many puzzling anomalies which emerge when historical timescales (i.e., those based on recorded history) are compared with dendrochronological and radiocarbon ages. Nowadays it is claimed more and more frequently that radiocarbon ages must never be referred to as true solar-year ages. It may well be that, when the other radiometric dating techniques are subjected to the same close scrutiny as radiocarbon dating, similar and even more severe problems may be encountered, leading to yet more adjustments of the timescale of the past.

As described in Chapter 2, astronomers have long thought that the Earth's rate of rotation must be slowly decreasing, and calculations have shown that the length of the day increases by about two seconds every 100,000 years. During the Triassic period there were about 382 days per year. During the Ordovician period there were 410; and at the beginning of the Cambrian period there were 421, with each day lasting for just 21 hours (one can't take this too far, of course, since otherwise a day at the time of formation of the Earth would have lasted just 18 minutes). If the length of a day on planet Earth has varied through geological time, why not the length of a year? And could there not have been occasional disruptions in geological time such as those preached by Immanuel Velikovsky in his books *Worlds in Collision, Ages in Chaos* and *Earth in Upheaval*? Most Earth scientists prefer to ignore Velikovsky and other latter-day catastrophists, but, although their theories are packed with errors, we cannot afford to assume that the problems of geological time have been solved. In spite of the convincing welter of radiometric dates now at our disposal, we still cannot claim to have a scale of absolute time. Time is something which geologists still do not properly understand, even though they may have a more realistic view of time than most.

BSJ

7 Time in Disarray

It was the late "Professor" Joad who, in his *Guide to Modern Thought*, used the phrase "the undoubted queerness of time". He was speaking about that curious and still warmly debated case of the two English ladies at Versailles in 1901, who apparently experienced a "time slip" and found themselves back in the Versailles of 1789, just before the downfall of the king. Both women felt depressed, and experienced a dream-like sensation; but neither realized that anything unusual had taken place until they compared notes later, and decided that it *was* rather odd that the Trianon park should have been visited that afternoon by so many people in period costume. Their book, *An Adventure* (1911), excited widespread attention because it was so obvious that the ladies—principals of an Oxford college—were of unquestioned integrity.

In 1965, the late Philippe Jullian published a biography of the French dandy Robert de Montesquiou (Proust's Baron Charlus) which at first seemed to explain the whole strange story. It seems that in the 1890s Montesquiou, Mme de Greffuhle and other members of Paris Society organized a fancy dress party in the dairy at Versailles, and spent some time before the party rehearsing theatricals in period costume. Dame Joan Evans, the literary executor of the two ladies—Charlotte Moberly and Eleanor Jourdain—was so convinced that her aunt had merely stumbled into a rehearsal that she decided to stop the reprinting of *An Adventure*. Her decision was undoubtedly premature, for Montesquiou's fancy dress party took place in 1894, seven years too early for the Moberly-Jourdain visit. In any case, a letter from Madame de Greffuhle reveals that she was in London on the day the "adventure" took place. So the mystery remains unexplained.

Joad concludes: "While admitting that the hypothesis of the present existence of the past is beset with difficulties of a metaphysical character . . . I think that it indicates the most fruitful basis for the investigation of these intriguing experiences." What exactly

Eleanor Jourdain (*left*) and Charlotte Moberly (*right*).

did he mean by "the present existence of the past"? He never
bothered to explain. But the phrase seems to suggest a notion that
is not too difficult to grasp: that the past is somehow alive and still
among us, like the voice of Caruso preserved on gramophone
records. In fact, a similar hypothesis was advanced in the middle of
the last century by Dr Joseph Rodes Buchanan, a professor of
medicine who came to believe that all physical objects carry their
history somehow imprinted on them—almost in the manner of a
photograph—and that this history can be "read" by a person
sensitive enough to pick up its vibrations. He called this ability
"psychometry", and the word has entered the vocabulary of paranormal research. Buchanan's brother-in-law, William Denton—a
professor of geology at Boston—was immensely excited when some
of his experimental subjects were able to describe in detail the
history of various geological samples—wrapped in brown paper—
and stated his conviction that this faculty would one day provide a
kind of telescope by means of which we would be able to see into
the past.

But then Buchanan's "psychometry" was not literally the ability
to *see* into the past—any more than a gramophone stylus is a time

machine that can transport you back into the life of Caruso. If the faculty exists—and there is much convincing evidence that it does—then it could be explained simply as a very highly developed ability to "read" the history of objects, rather as Sherlock Holmes was able to tell Watson the history of his alcoholic brother from the evidence of his watch. And this, I suspect, is not precisely what Joad meant by the "undoubted queerness of time". For, in the section before his account of the "adventure" of Miss Moberly and Miss Jourdain, he discusses J.W. Dunne's book *An Experiment with Time*; and Dunne's book is an account of how he had certain clear and detailed dreams of the *future*. If Dunne's book is to be believed—and, again, he had a reputation for integrity—then he dreamed of such events as the great Martinique earthquake some weeks before it happened. And this is utterly unexplainable on any "scientific" theory of time, no matter how abstract and complex: as we saw in Chapter 5, the scientists' view of time dictates that the future cannot affect the past. I may be able to explain certain personal premonitions—say, the death of a relative—in logical terms (i.e., I knew he was ill and suffered from a bad heart), but to

J.W. Dunne (1875–1949), the British philosopher whose *An Experiment with Time* profoundly altered our ideas of perceived time and even the nature of time itself.

dream of a volcanic explosion on an island you know nothing about is obviously an event of a different order.

There, then, is the problem. The files of the Society for Psychical Research are full of convincing cases of premonitions of the future and of "prophecy". And they flatly contradict everything that human beings know—intuitively—about time. The one thing that is absolutely certain about our world is that everything that is born ends eventually by dying, and that, in between these two events, it gets steadily older. Time is irreversible. With the aid of a tape recorder, I can replay the voice of someone who is dead; but, if I happen to feel guilty about the way I have treated him, there is absolutely no way in which I can go back in time and "unhappen" what has happened. We all know this. It is not only a fundamental part of our experience; it seems to be a law of the Universe.

Now when, in 1895, H.G. Wells wrote his science-fiction story *The Time Machine* he introduced his readers to an exciting and fascinating new hypothesis. Time, says Wells' Time Traveller, is nothing more than a fourth dimension of space. Consider photographs of a man at the ages of eight, fifteen, seventeen, twenty-three, and so on. These are basically three-dimensional representations of a four-dimensional being, rather as you might take slices or cross-sections of a length of soft clay. What this implies is that each cross-section is in some way false or, at least, misleading—exactly as those flat Egyptian portraits of solid human beings are misleading. Seen from the perspective of the fourth dimension, a man is a single chunk of matter stretching from one point in time to another, not a three-dimensional chunk of matter *moving* from one moment to the next.

One of the Time Traveller's companions objects that we cannot move about in time; whereupon he makes an interesting reply: "You are wrong to say that we cannot move about in Time. For instance, if I am recalling an incident very vividly I go back to the instant of its occurrence: I become absent-minded, as you say. I jump back for a moment. Of course, we have no means of staying back for any length of Time, any more than a savage or animal has of staying six feet above the ground. But a civilized man is better off than the savage in this respect. He can go up against gravitation in a balloon, and why should he not hope that ultimately he may even be able to stop or accelerate his drift along the Time-Dimension, or even turn about and travel the other way. . . ?"

The Traveller, of course, claims to have invented a machine for doing precisely this. But the interesting point of the above expla-

The Time Traveller sets off on his epic voyage . . . A still from the movie of H.G. Wells' *The Time Machine*.

nation is that it suggests a quite different method of time travel. Wells says that when we recall an event vividly, we move back into the past for a moment; but we have no capacity to stay there. Time, he says, in another paragraph, is essentially *mental* travel from the cradle to the grave. What Wells is suggesting is that time travel is a mental faculty we already possess, but to a very slight extent.

Wells himself apparently forgot that important suggestion, thrown off casually in the opening chapter of *The Time Machine*. And the remainder of his story—with its mechanical flight through time—raises the kind of paradoxical questions that have become a commonplace of science fiction ever since. For example, as he moves into the future, he sees his housekeeper come into the room and move across it with the speed of a bullet: for now he is moving more swiftly through time, her action happens in a shorter space of time. If he had been going backwards in time, he would have seen her move across the room backwards, her actions reversed. But then, would he not also have seen *himself,* as he was a few minutes before, or the day or month before? In fact, what was to prevent him halting the Time Machine and going to shake hands with his "self" of yesterday? Or why should he not go forward to his self of tomorrow and ask him what horse won the Grand National? He could even ask his self of tomorrow and his self of yesterday to climb into the Time Machine and accompany him back to today for dinner . . .

And already we see the emergence of the paradox. What right has the Time Traveller to regard his own time as *the* present, and his own "self" as *the* Time Traveller? Wells sidesteps this question by sending the Traveller backwards or forwards in time *beyond* his own life span. So if he went back to 1812 to meet Napoleon or 1066 to meet King Harold, it would *sound* perfectly logical, if unbelievable. But if the Time Traveller consists of millions of "selves", one for every split second of his life, then the same goes for every other person and object in the Universe. The trouble with this is that every one of these multiple beings would have its own past and future, since each is a separate individual. (For example, if the Time Traveller invited his selves of yesterday and tomorrow for dinner, each would proceed to travel into the future separately as three separate beings.) You end up with an absurd vision of a multiple-multiple Universe in which everyone is fragmented into an infinite number of selves . . .

It is, of course, mere fiction, so we can forgive its shortcomings. But then, the actual experience of time travel is *not* mere fiction. I suggested, for example, that the Time Traveller of today might pay a call on his self of tomorrow to enquire the winner of the Grand National; he could then go back to his own time and place a large bet on it . . . But such events have, in fact, occurred. In 1976 I made a television programme for BBC2 about John Godley, Lord Kilbracken, who, as an Oxford undergraduate, dreamed winners of horse races, and made several useful sums of money through his curious ability. Peter Fairley, the science correspondent of Independent Television, had a similar experience. In a BBC broadcast, he told how, as he was driving to work one day in 1965, he heard a request on the car radio for a Mrs Blakeney; he had just driven through the village of Blakeney, and a few minutes later, heard a reference to another—totally unconnected—Blakeney. At the office he heard the name again, this time a horse running in the Derby. He backed it and it won. From then on, he explained, he could pick winners merely by looking down a list of horses; the winner would "leap off the page" at him. He said that as soon as he began to think about it and worry about it, the faculty vanished . . .

Now this is altogether closer to Wells' suggestion of Time Travel as a purely mental faculty. And it is certainly far more convincing than the version involving time machines.

Time and Mind

This is the point where, I feel, the reader should be prepared to ask himself a question. Does he consider this discussion as amusing but purely academic? Or is he, in fact, prepared to believe that people like Dunne, Godley and Fairley are telling the truth, that therefore time *is* much queerer than we are willing to acknowledge? I suspect that most readers, even the open-minded ones, are really, deep down, thoroughgoing sceptics, who feel that time is very much what we suppose it to be: a one-way street, and that therefore stories about glimpses of the past and future should be treated as interesting but amusing speculations, like the writings of Erich von Däniken.

If, on the other hand, you are willing to face up to the possibility that time may be more paradoxical than our commonsense view of it, then you have taken a very remarkable step indeed—perhaps one of the most daring imaginative leaps of a lifetime. For even the most vital and self-confident human beings feel that, in a basic sense, they are trapped in time. There are many problems that we *can* solve, many difficulties we can evade, many confusions we can untangle; but ultimately, it seems, we are all slaves of Father Time. This notion is so deeply ingrained that we take it utterly for granted. Yet *if* it were untrue, it would suggest that life is not at all what we suppose it to be, and that our poor-spirited notion of ourselves is an insult to our true possibilities.

Assuming, then, that we are willing to give serious consideration to this possibility that time may be queerer than we realize, we find ourselves faced with some strange and interesting consequences.

Let us, to begin with, agree that the usual notion of time travel, derived from Wells, is absurd and self-contradictory. In *that* sense, the past is the past and the future is the future, and we can never hope to explore either with the aid of a Time Machine. For in this sense, time does not exist; it is a semantic misunderstanding. I tried to explain the reason for this in a passage of my book *The Occult*. Suppose people were born on moving trains and stayed on them until they died. They might invent a word to describe the everyday sensation of scenery flowing past the window, a word like "zyme". When the train stops in stations they would say that zyme has halted; if the train reverses, they would say zyme is flowing backwards. But if someone spoke of zyme as an entity, they would obviously be committing a logical error; it consists of *many* things—a railway carriage, scenery, motion and so on. The same goes for time. It is basically a *process* which involves physical objects. If you think of a completely "empty" Universe, or a completely static Universe, it would obviously have no time. *This* is why Wells' time machine is an absurdity.

If Peter Fairley could really predict which horse would win a race, then there is clearly something wrong with our human notion of time; for the idea that the future has already taken place—which it must have done if you are to "know" it—is self-contradictory, a paradox. But then, our minds are a paradox in precisely the same sense. You and I apparently exist in a solid, three-dimensional Universe; we are physical objects. Then where, precisely, is my mind? Inside my head? "Realist" philosophers have tried hard to explain mind in physical terms—the brain and the nervous system—but they end with a static model, rather like a computer. And a computer needs to be *worked* by somebody. When I struggle with an intellectual or emotional problem, I am aware of an element that I call "me" trying to get the best out of the computer. This being can look on quite detachedly while "I" am flooded with a powerful emotion. It applies the accelerator or brake to my moods and feelings. It seems to exist in a dimension apart from this physical world we live in.

To me, these considerations suggest that these two paradoxical concepts—time and the mind—are closely connected. Our bodies exist in the realm of one-way time, but our minds do not. As Wells points out, when I become absent-minded, my mind goes "elsewhere". But on the whole, these visits to other times and places are far less vivid than our everyday lives. Yet this is not so much a limitation of our minds as of the "computer" they use, the brain.

For example, there is an important experience of the philosopher J.B. Bennett (which I have cited elsewhere), described in his autobiography *Witness*. Bennett tells how, when he was staying at the Gurdjieff Institute at Fontainebleau, he woke up one morning feeling exceptionally weak from dysentery, but nevertheless forced himself to get up. Later that morning he took part in some Gurdjieff exercises—incredibly difficult and complex physical movements. One by one, the other disciples dropped out; but, in spite of extreme fatigue and discomfort, Bennett forced himself to go on.

Then, quite suddenly, "I was filled with an influx of an immense power. My body seemed to have turned into light." All fatigue vanished. When he went outside, he decided to test this power by digging at a rate he could not ordinarily maintain for more than a few minutes; he was able to continue for half an hour without fatigue. He walked out into the forest, and decided to try to test his control over his emotions. He willed himself to feel astonishment. "Instantly, I was overwhelmed with amazement, not only at my own state, but at everything I looked at or thought of." The thought of "fear" filled him with immense dread; the thought of "joy" filled him with rapture; the thought of "love" flooded him with a tremendous tenderness and compassion. Finally, bewildered by this new ability to feel anything he liked, he willed it to go away, and it instantly vanished.

Now what is involved here is obviously what William James calls "vital reserves". James points out that we can feel exhausted, push ourselves *beyond* the exhaustion, and suddenly feel full of energy again. It is the phenomenon of "second wind". It seems that we possess vast energy reserves that we fail to make use of. But a sudden emergency will bring them into operation. Bennett's tremendous effort not to drop out of the Gurdjieff exercises somehow pushed him into a heightened state of "second wind", and brought a completely new level of control over his "computer". It is a pity that he did not try the experiment of recalling some event from his past; I suspect that he would have been able to "replay" it in the most accurate detail.

In fact, as Dr Wilder Penfield discovered, our brains contain the stored "memory tapes" of everything we have ever seen or felt, and these tapes can be "replayed" by stimulating the temporal cortex of the brain with an electric probe. If we could achieve Bennett's state of "second wind", the electric probe would be unnecessary; all the memory tapes of the brain would become instantly accessible to us . . .

But that, you will object, is still not time travel; it is merely playing back a recording. True. But, if Joseph Rodes Buchanan and William Denton were correct about "psychometry", then the brain also has the power to play back the history of any object it chooses to scan—for example, a five-billion-year-old meteorite. Buchanan's "sensitives" could hold a sealed letter and describe not only its contents but also the state of mind of the person who had written it. And this, you may point out, is still not time travel. True. But it is something very like it. And I would remind you that we have already agreed that time travel, in Wells' sense, is an absurdity. You cannot literally go back "before" the Battle of Hastings, because the Battle of Hastings has already happened, and it cannot be unhappened. Yet, if Buchanan and Denton are correct, then it should be possible for a "sensitive" to literally relive a day in the life of a soldier who fought at the battle of Hastings. And Dunne's experiment with time seems to suggest that it might be possible to do the same for the future, and "relive" a day that has not yet taken place. And this, I think, *would* qualify as time travel.

What I am now suggesting is a view of the human mind that has been forcing itself upon me for many years. My starting point, in books like *The Outsider* and *Religion and the Rebel,* was the experiences of certain poets and mystics. The romantic poets of the nineteenth century seemed to differ from their predecessors in one important respect: they seemed to have an altogether greater capacity for sustaining *imaginative intensity.* We live our lives confined by space and time and the trivial necessities of everyday life; consciousness is basically a device for perceiving what goes on around us. Poets and mystics seem to be able to use it for a quite different purpose—to build up a kind of internal world whose intensity rivals that of the physical reality that surrounds us. When I came—almost by accident—to turn my attention to the realm of the "occult" or paranormal, it struck me that the "psychic" is only another type of poet: a person for whom the physical world is only one aspect of reality.

Now this view seems to me, on reflection, logical and reasonable enough. Consciousness is tied to the physical world for a simple reason: if it weren't, we would have been extinct long ago. As H.G. Wells pointed out, all animals are "up against it" from the moment they are born. In the Victorian age, children began work at six in the morning and finished at eight in the evening. Life is still brutal and hard for well over a half of the human race. *I* am lucky that I can sit at my desk, in a comfortable room, and address my mind to this interesting problem of the nature of time; you are lucky that you can sit down and read it. If you and I had to work a fourteen-hour day in a factory we would long for a little leisure to relax and allow the mind to wing its way through the worlds of imagination.

Because of this harsh physical necessity, consciousness has accustomed itself to sticking to the material world: which means, in effect, that it has never had a chance to explore its own capacities—or rather, the capacities of that extraordinary computer called the brain. But here we come to one of the strangest parts of the story. For some odd reason, the capacity of this computer is far greater than it needs to be—at least, in terms of Darwinian evolution. For example, it is quite clear that we never make use of that vast library of "memory tapes" that Wilder Penfield discovered; we don't *need* to make use of them for everyday survival. Then why are they there? Why has evolution dictated that the brain should remember every tiny event and idea of our lives? Again, I have always been fascinated by the capacity of calculating prodigies—usually young children of ordinary intelligence—who can multiply or divide immense sums in their heads. Equally extraordinary is the class known as "idiot savants"—children whose IQ may be on the moron level, yet who, in one particular field, have some incredible mental gift—one, for example, could reel off the name of every musical film ever made and every actor who played every part. Moreover, some of these idiot savants have highly developed "psychic" powers; for example, one boy declined a lift home with his teacher because, he said, his mother would be meeting him out of school. In fact, his mother *did* arrive to meet him; but she had decided to do so only half an hour before, when another trip took her close to the school . . .

And this example brings me to the starting point of my book *The Occult*: the observation that "psychic powers" often seem to involve a breakdown—or at least, loss of efficiency—in our normal mental powers. For example, a Dutch house painter named Peter van der Hurk fell off his ladder and fractured his skull; when he woke up in hospital, he discovered that he "knew" all kinds of things about his fellow patients, about their past and even their future. This strange capacity has remained with him and, under the name of Peter Hurkos, he has made a considerable reputation as a "clairvoyant" and psychometrist, often helping the police to solve murder cases. But, in the days immediately following his accident, he found life difficult because his new psychic powers made it impossible for him to concentrate on ordinary, everyday jobs; he might have starved if someone had not suggested using his powers to make a living as a stage "magician". When I read this story in Hurkos' autobiography I found myself thinking of all those romantic poets and artists who had died in poverty because they found it impossible to concentrate on the dreary necessities of material existence. There is obviously a close analogy.

All this seems to suggest that our brains possess extraordinary powers that most of us never have reason to use. The problem of survival demands that we are tied down to the everyday world; if this were not so, we might all be calculating prodigies and psychics, and probably literary and artistic geniuses into the bargain.

But to phrase it this way suggests that it is a question of either/or; either we get rid of such unusual faculties or we lose our ability to survive. But is the choice really as harsh as that? I am inclined to doubt it. Life for most of us is safer and more secure than at any other time in history. Modern man is far less likely to be knocked down by a car than his ancestors were to be eaten by wild beasts or killed by their fellow men. (Even as recently as the age of Dr Johnson, remote country houses were often besieged by gangs of ruffians who killed those who resisted and carried off everything of value.) Most of us have hours of leisure every week in which we might explore the possibilities of human consciousness. No, the real problem is a force of habit so deeply ingrained that it would be better to refer to it as hypnosis. If you force a chicken's beak against the floor, then draw a chalk line straight in front of it, the chicken will be unable to raise its head when you let it go; for some odd reason it focuses attention on the chalk line, and becomes hypnotized by it. We all suffer from a similar tendency; the moment we relax, habit induces a state similar to hypnosis, in which the attention becomes fixed on the external world. Sartre wrote about the café proprietor in *Nausea*: "When his café empties, his head empties too." But it is not confined to the illiterate or unintelligent. There is a story told of the famous mathematician Hilbert. Before a dinner party, his wife sent him upstairs to change his tie; when, after an hour, he had still not reappeared, she went to see what had happened; he was in bed fast asleep. He explained that as soon as he had removed his tie, he had automatically taken off the rest of his clothes, put on his pyjamas and climbed into bed.

This is the problem of human consciousness: habits that bundle us into bed and off to sleep when there are far more interesting things to be done. Chesterton asked why the world is so full of bright children and dud grown ups. The reason is that our most interesting potentialities fail to survive adolescence; we slip into a habit of using only a fraction of our powers.

When habit is broken, anything can happen. In a book called *Mysteries* (1978) I have cited the case of a lady named Jane O'Neill who, when driving to London airport, witnessed a serious accident and helped to free badly injured people from a wrecked coach. The

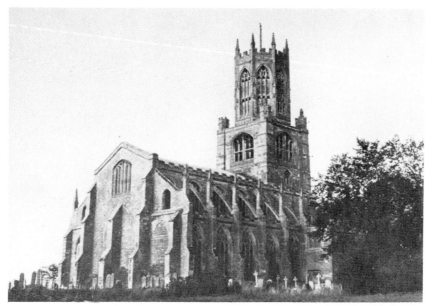

Fotheringhay Church, where Jane O'Neill had such an unusual brush with time in disarray.

shock was so severe that she had to take several weeks off from work. She began to experience strange waking visions, some of which were oddly accurate: for example, she "saw" a close friend chained in the galleys; told about this, her friend replied that her ancestors were Huguenots and many *had* found themselves in the galleys. One day in Fotheringhay Church, Jane O'Neill was impressed by a picture behind the altar. She later mentioned this to the friend who had accompanied her, and her friend said that *she* had not seen any picture. Miss O'Neill was so puzzled that she rang the lady who cleaned the church and asked her about it; the lady replied that there was no such picture. Later, the two women revisited the church; to Jane O'Neill's surprise, the inside was quite different from what she had seen before—it was much smaller—and the picture was not there. She asked an expert on East Anglian churches, who put her in touch with a historian who knew the history of Fotheringhay. He was able to tell her that the church she had "seen" had been the church as it was more than four centuries ago; it had been rebuilt in 1553 . . .

Jane O'Neill's experience is, in its way, as well authenticated as that of Miss Moberly and Miss Jourdain. In one sense, it is more convincing; I heard of it by accident, through a friend, and wrote to Miss O'Neill, who was kind enough to send me a full account,

together with the exchange of letters with the historian which established that she had "seen" the earlier church. Miss O'Neill had
made no attempt to publish her interesting story, so cannot be
accused of attention-seeking.

But how can we reconcile a story as extraordinary as this with
our everyday experience of the real world? Most scientists have a
short and convenient method of dealing with such anomalies; they
dismiss them as lies, distortions or mistakes. Whether intellectually
justified or not (on grounds of "the laws of probability"), this is
bound to strike anyone interested in such matters as pure mental
laziness. If an answer is to be found, I believe that its starting point
must be the notion that the powers of the human mind are far less
limited than we naturally assume. This was a conclusion I had
reached many years before I became interested in the paranormal;
so that, for example, in *Religion and the Rebel* (1957), I had suggested
that our everyday consciousness is as limited as the middle few
notes of a piano-keyboard, and that its possible range is as wide as
the whole keyboard. In states of great happiness or relief, or when
involved in some absorbing adventure, we receive a clear intuition
that the world is an infinitely richer and more complex place than
ordinary consciousness permits us to perceive. And, moreover, that
the mind is perfectly capable of taking a wider grip on that breadth
and complexity . . .

Hurkos' accident, like Jane O'Neill's, shook his mind out of its
usual narrow rut, and made him aware that "everyday consciousness" is basically unreliable in its report about the actuality that
surrounds us. But then, is not such narrowness preferable to the
state of confused inefficiency that accompanied his powers of
"second sight"? Was Jane O'Neill's glimpse of Fotheringhay in the
sixteenth century (or earlier) *worth* the mental shock of the coach
accident? These questions raise serious doubts about the desirabil-
ity of such powers. But then, we are assuming that it is possible to
investigate the unknown powers of the mind only by destroying our
everyday sense of reality. And this, fortunately, is untrue.

This is illustrated by a story told by Alan Vaughan in a remarkable work, *Patterns of Prophecy*, a study of the scientific evidence for
precognition. He explains how he became interested in precognition. In 1965, then a science editor, he purchased a Ouija board
to amuse a sick friend, and was surprised by the accuracy of some
of its "information". An entity who called himself "Z" seemed
particularly accurate. As he continued to experiment, Vaughan
received messages from a neurotic personality who called herself

Nostradamus (1503–1566), the French astrologer whose predictions have made him the most famous seer of all times.

Jeanne Dixon, the US clairvoyant who has startled many by her predictions of such events as the assassination of J.F. Kennedy; it must be remembered, too, that many of her predictions have certainly not come to pass.

"Nada", and who claimed to be the wife of a Nantucket sea captain; she seemed jealous that Vaughan was alive while she was dead. Then "Nada" somehow got inside his head and he found himself unable to get rid of her. Asked what was happening, "Z" replied that it was a case of possession. "Z" made Vaughan write out the sentence: "Each of us has a spirit while living. Do not meddle with the spirits of the dead."

"As I wrote out this passage, I began to feel an energy rising up within my body and entering my brain. It pushed out both 'Nada' and 'Z'. My friends noted that my face, which had been white and pinched, suddenly flooded with colour. I felt a tremendous sense of elation and physical wellbeing. The energy grew stronger and seemed to extend beyond my body. My mind seemed to race in some extended dimension that knew no confines of time or space. For the first time, I began to sense what was going on in other people's minds and—to my astonishment—I began to sense the future through some kind of extended awareness. My first act in this strange but exciting state was to throw the Ouija down an incinerator chute . . ."

It was this experience that led Vaughan to study the whole question of prophetic glimpses of the future. He had *seen* this "extended dimension that knew no confines of time or space", and decided that it deserved to be investigated. The poet Robert Graves described a similar experience in a story called "The Abominable Mr Gunn" (which, he told me, was autobiographical). "One fine summer evening as I sat alone on the roller behind the cricket pavilion, with nothing in my head, I received a celestial illumination: it occurred to me that I knew everything. I remember letting my mind range rapidly over all its familiar subjects of knowledge; only to find that this was no foolish fancy. I did know everything. To be plain: though conscious of having come less than a third of the way along the path of formal education . . . I nevertheless held the key of truth in my hand, and could use it to open any lock of any door. Mine was no religious or philosophical theory, but a simple method of looking sideways at disorderly facts so as to make perfect sense of them."

The "secret", Graves says, was still there when he woke up the next morning; but, when he tried writing it down, it vanished.

It is true that Graves fails to explain just what he meant by the "secret", except to say that it was "a sudden infantile awareness of the power of intuition, the supra-logic that cuts out all routine processes of thought and leaps straight from problem to answer". But he offers a further clue in citing the case of another boy in the school who was able to solve a highly complicated arithmetical problem merely by looking at it. The form master—"Mr Gunn"—accused the boy of looking at the answer at the end of the book; the boy replied that he *had* checked with the answer—later—and that its last two figures were wrong—they should be 35, not 53. The unsympathetic and obtuse Mr Gunn sent the boy to the head-master for a caning, declining to believe that he could simply have "seen" the answer . . .

So it seems that Graves is speaking of a power related to that of mathematical prodigies, the ability of the mind to *see* the answer to a problem in a single flash. And how, precisely, does such an ability work? Is it some form of lightning calculation, that is, a process of ordinary reason in which everything is speeded up, as in the famous Trachtenberg speed system of mathematics? Apparently not. We know this from the case of Zerah Colburn, the Canadian calculating prodigy, who was asked whether a certain immense number was a prime (i.e., could not be divided by any other number), and who replied instantly: No, it can be divided by 641. Now there is no mathematical method of determining whether a certain number is a prime—except the painful method of trial and error, dividing it by every smaller number and deciding that none of them works (shortcuts exist: if it can't be divided by 3 it can't be divided by 6, 9, 12, 15 . . .). Obviously, Colburn "saw" the answer, like Graves' fellow pupil F.F. Smilley did—from "above", as it were: a kind of bird's eye view. And Graves' "secret" was, presumably, some similar method of grasping the answer to any problem by instantaneous intuition . . .

The Divided Brain

Now at this point it is worth mentioning a recent discovery in brain physiology: the recognition by R.W. Sperry that we have virtually two people living inside our heads, in the right and left cerebral hemispheres. We have known for a long time that the left half of the brain governs language, while the right is concerned with recognition. The left is concerned also with logic and reason, while the right deals in "appreciation"—for example, artistic enjoyment. You could say that the left is a scientist, while the right is an artist.

The two halves of the brain are connected by a bridge of nerve fibre. If this is severed, they proceed to work separately. So if the left half of the brain (which is, in fact, connected to the right eye) is shown an apple, and the other half is shown an orange, and you are asked what you have just seen, you reply: an apple. But if you are asked to *write down* with the left hand what you have just seen, you will write: an orange. If you are asked what you have just written, you reply: an apple. Neither half knows what the other is thinking.

But the most significant insight arising from this experiment is that the being you call "you"—your ego—resides in the left half of your brain. There is another "you" a few inches away, in the right half; but it is dumb.

When I work out a sum on paper, I am using my left hemisphere—with a certain amount of occasional assistance from the right, by way of sudden insights. And this, on the whole, seems to be the way the human brain works: the left is the "front man", the ego that deals with the world; and the right has to express itself *via* the left. And, on the whole, the right has a fairly hard time of it; for the left is always in a hurry, always working out problems, and it tends to treat the right with impatience. This is why civilized Man seems to possess so little intuition.

It seems probable that calculating prodigies have not yet fallen victim to this bullying dominance of the left. The "shades of the prison house" have not yet begun to close. They *see* the answer to a problem, and pass it on instantaneously, unimpeded by the usual red tape of the bureaucrat who lives in the left brain.

For this, I must stress, is the real problem of civilized Man. We have evolved to our present level through the use of language and concepts. We use these so constantly that we "identify" with the left half of the brain. This does no real harm, for in a sense, the "personality" *is* the linguistic part of us. The trouble arises from the *attitude* of the ego to the non-ego who lives in the right cerebral hemisphere. We tend to treat it as an idiot, as a kind of inarticulate and not-very-bright younger brother who is always being ignored and told to shut up. If we took the trouble to listen to it, we might learn a great deal. Occasionally, it may become so alarmed at our carefully calculated stupidities that it takes the law into its own hands and interferes. Here I can cite a personal example. The hill that leads up from Pentewan to Mevagissey is long, and has several abrupt curves. One day, I was driving up this hill with the Sun in my eyes, almost completely blinded. At a certain point I reasoned that I must be approaching a bend, and tried to turn the steering wheel. *My hands ignored me*; they kept the wheel steady. My right brain knew I had not yet reached the bend, and simply cancelled my order to turn the steering wheel.

Even this last sentence illustrates our basic mistake. I say "*my* hands", "*my* right brain", as if they were both my property, like my clothes. But the being who calls himself "I" is a usurper. It is his brother, who lives next door, who is, the rightful heir to the throne. I say this because the left, for all its naïve egoism, cannot live without the intuitions and insights of the right—there are many creatures in the world who live perfectly well without language or ideas. But the ideal state is one of close cooperation between the two halves, with the left treating the right as a wise counsellor and trusted adviser, not as the village idiot.

Significantly, the left brain has a strong sense of time; the right has absolutely none. It strolls along at its own pace, with its hands in its pockets. This does not mean that the right lacks the ability to calculate time—on the contrary, when you tell yourself that you must wake up at six o'clock precisely and you open your eyes on the stroke of six, this is the work of the right. But it declines to take time too seriously. And it is right to feel sceptical. The left is stupidly obsessed by time. An anecdote told by William Seabrook of Aleister Crowley illustrates the point. When Crowley was on the island of Cephalù, a film star named Elizabeth Fox came to pay him a visit; she was in a state of permanent nervous tension. Crowley told her that she must begin her cure with a month of meditation on the cliff top. The idea dismayed her, but she agreed. She lived in a lean-to shelter and a boy brought up water, bread and grapes every day at dusk. For the first few days she was bored and irritable. By the nineteenth day she felt nothing but boredom. Then, quite suddenly, she passed into a state of deep calm and peace, with no desire to move. What had happened was simply that her over-dominant left brain—accustomed to the Hollywood rat race—had gradually realised that it could stop running; then the right took over, with its sense of timelessness and serenity.

Faculty X and Insight

What is being suggested is that *time is an invention of the left brain*. Time, as such, does not exist in nature. Nature knows only what Whitehead calls "process"—things happening. What human beings call time is a psychological concept; moreover, it is a left-brain concept.

Now the left brain, as we know, sees things in rigid categories, and nature does not operate within such categories. Consider Zeno's paradox of the arrow. At any moment it is either where it is or where it isn't. It *can't* be where it isn't; but if it is where it *is*, then it can't be moving. The paradox of Achilles and the tortoise depends on the same kind of logic. But the arrow *does* move; Achilles *does* overtake the tortoise, although it is "logically" impossible. According to the left brain, there is no logical way of deciding whether a large number is a prime except by trial and error; but Zerah Colburn's right brain solved it instantly; and, in the same way, Peter Fairley's right brain knew in advance which horses would win at the races. (Significantly, Fairley had suffered temporary blindness just before he developed this ability; it seems probable that the shock was responsible for "short-circuiting" the usual left brain processes.)

This essay is, of course, written in language, and it makes use of concepts; consequently its aim is, to some extent, self-defeating. How can I convey in words the notion that time itself is merely a concept? The above examples can at least take us in the right direction. For most people have known what it is to suddenly "know" the answer to a problem without thinking it out. Everyone has had the experience of trying hard to remember something, and then having it stroll into his brain when he was no longer trying— almost as if another person had knocked on the door of the left brain and said: "Is this what you were looking for?"

Which brings me to the most important step in this argument: that everyone has experienced the most basic "right-brain" insight, the curious ability that in *The Occult* I labelled "Faculty X". This is simply that odd ability to suddenly grasp the *reality* of some other time or other place. I have elsewhere cited the example of the experience that led Arnold Toynbee to begin his *Study of History*. Toynbee was sitting at the summit of the citadel of Mistrà, in Sparta, looking at the ruins that had been left by the wild highlanders who had overwhelmed it in 1821, when he was suddenly struck by the *reality* of what had happened—as if the highlanders were, at that very minute, pouring over the horizon and overwhelming the city. He goes on to describe half a dozen more occasions when the "historical imagination" has suddenly "brought the past to life" and made it real, and ends by describing a semi-mystical experience that occurred as he was passing Victoria Station, London, during World War I, when he found himself "in communion, not just with this or that episode in History, but with all that had been, and was, and was to come".

Chesterton once said: "We say thank you when someone passes us the salt, but we don't *mean* it. We say the Earth is round, but we don't mean it, even though it's true." We mean something only when we feel it intensely, here and now. And this is what happens in flashes of Faculty X: the mind suddenly conjures up the *reality* of some other time and place, as Proust's hero suddenly became aware of the reality of his childhood as he tasted the cake dipped in herb tea.

Faculty X is another name for insight, the sudden flash of understanding, of direct knowledge. And it enables us to see precisely how the left and right cooperate. At school, I may learn some mathematical formula, like those for doing long division or extracting square roots; but I use it mechanically. If one day I forget the formula, and have to work it out for myself, I achieve insight into the reasons that lay behind it. But I can quite easily forget this insight, and go back to a mechanical use of the formula. The left brain deals with surfaces, with forms; the right brain deals with insights, with what lies beneath the surface. The left brain is a labour-saving device, an energy-saving device—exactly like using some simple mnemonic to remember the colours of the spectrum or the black notes on the piano. It is when you are full of energy— perhaps on a spring morning—that the right brain produces that odd glowing sense of reality. When you are very tired, the left brain takes over. Constant mental fatigue can produce the state Sartre calls "nausea", in which the left brain scans the world but lacks all insight into its meaning—the right has gone off duty: reality seems crude and meaningless.

But here is the most difficult part of the argument to grasp. It is the right brain which presents us with "reality". The left presents us only with *immediacy*, what happens to be here and now. The left "scans" the world; the right adds meaning and value. And your eyes, which are now scanning these words, are actually *telling you lies*. For they are presenting an essentially unreal world to you as the only reality. "This is real," I say, knocking on the table with my knuckles; but my knuckles are only scanners, like my eyes.

If, as you read these lines, you can penetrate to the meaning I am trying to convey, you will do it by a mental *leap*, from left to right. And if you can make that leap, you will also be able to grasp how Peter Fairley could know the winners of a race that had not yet taken place, or how Zerah Colburn could "know" that 4,294,967,297 is divisible by 641. Somehow, the right "thinks" vertically, by taking a kind of upward leap and simply looking down on the answer. You will object that this still doesn't explain how it could "look down on" the future, but this is because you are still thinking in left-brain terms. How would you, in fact, go about predicting some future event, assuming that someone made it worth your while to do so? You would ploddingly try to assemble thousands of present "trends", and try to work them out according to the law of probabilities. And because there are so many billions of possibilities, we say the future is unpredictable. The right brain appears to know better . . .

Let me try to summarize the argument so far. We have begun by dismissing "time" in the Wellsian sense, the kind of time in which you could travel with the aid of a time machine. Like "zyme", this time is a logical error. What really happens out there is "process", and it would be absurd to speak of travelling in process. Time is actually a clock ticking inside the head—and, what is more, in only one side of the head. Our senses, which are built to "scan" the world, chop up process into seconds and minutes. They force us to see the world in these rigid terms of spatial and temporal location. Kant was quite right when he said that we see the world through "categories". Think of the Kantian categories as a weird pair of prismatic spectacles you wear on your nose, spectacles which turn everything you see into the strangest angles and corners. *This* is space and time, as our brains grasp it.

All this, of course, fails to answer a basic question: how future time—that is, process which has not yet taken place—can be predictable. The only scientific explanation is the one we have considered, the statistical assessment of "trends". But it seems fairly clear that Peter Fairley was unable to spot winners by this method, for he knew nothing about racing, let alone about the complex possibilities presented by all the horses in the race. Anyway, experiment has shown that this cannot be the explanation. The well-known psychical investigator S.G. Soal performed a series of experiments in telepathy with a man named Basil Shackleton, and both were disappointed that the results seemed to be negative. Then a careful look at the results revealed an interesting thing: Shackleton was guessing the *next* ESP card that would be chosen. This was confirmed by substituting cards with animal pictures— zebras, giraffes, and so on. Now there could be no possible doubt. If Soal uncovered a card of a zebra, and Shackleton (sitting in the next room) named it as a giraffe, it was almost certain that the *next* card Soal turned over would be a giraffe. Other experimenters— like J.B. Rhine and Charles Tart—have produced similar results.

So it looks as if we are faced with a basic fact: that, whether it is impossible or not, precognition actually takes place—precise and detailed precognition of the future—which suggests clearly that the "Kantian" theory is basically correct: there is something *wrong* with what our senses—and left brain—tell us about the world.

I could easily spend the remainder of this chapter raising questions about precisely how our senses could be mistaken. Such an approach would be interesting; but I doubt whether it would be very conclusive. Besides, much of my time would be taken up in summarizing Edmund Husserl's book *The Phenomenology of Internal Time Consciousness*; and those who are interested would do better to read it for themselves. Instead, let us, for the sake of argument, assume that this part of the case is proved—that there is something wrong with our left-brain conception of time—and look more closely into the other half of the equation: the curious power that, under certain circumstances, seems to enable us to foresee the future.

In a fascinating and lucid book, *The Case Against Jones*, John Vyvyan cites two interesting cases, one of precognition, one of retrocognition.

The first concerns a priest named Canon Guarnier, who dreamed with exceptional clarity of an Italian landscape—a mountain road, a white house, a woman knitting with her daughter looking on, three men dressed in aprons and pointed hats sitting at a table, a sleeping dog, three sheep in a field . . . The scene was detailed and vivid. Three years later, on his way to Rome, Guarnier's carriage stopped to change horses, and he found himself looking at the identical scene, accurate in every detail. "Nothing is changed; the people are exactly those I saw, as I saw them, doing the same things in the same attitudes, with the same gestures . . ."

The other case concerns the novelist George Gissing, who fell into a fever at Crotone in southern Italy. After a nightmare, he fell into a "visionary state", in which he saw a series of pictures of Roman history. These are described in considerable detail—too long to quote here. But Gissing himself had no doubt that he had somehow witnessed real scenes of history, not simply imaginative pictures. "If the picture corresponded to nothing real, tell me who can, by what power I reconstructed, to the last perfection of intimacy, a world known to me only in ruined fragments."

This, of course, is no proof that it was not imagination. What strikes me in reading Gissing's account—for example, of seeing Hannibal's slaughter of two thousand mercenaries on the seashore by Crotone—is its similarity to Toynbee's "visions" of the past. Wells' account of Gissing's death—in the *Experiment in Autobiography*—makes it clear that Gissing saw these visions again on his deathbed. Like John Vyvyan, I am certainly inclined to disbelieve that it was mere hallucination. His insistence on the clarity of the scene recalls Guarnier's dream, and the experiences of Jane O'Neill and of Misses Moberly and Jourdain.

I formulated the theory of Faculty X in my book *The Occult* (1971). But four years before this, I had made use of the concept in fiction, in a novel called *The Philosopher's Stone*, which is centrally concerned with this notion of "mental time travel". In this novel I suggested that the pre-frontal lobes of the brain (I didn't then know about the rôles of the right and left brains) are somehow connected with "poetic" experience: Wordsworth's feeling as a child that meadow, grove and stream were "apparelled in celestial light". No one seems certain of the precise purpose of the pre-frontal lobes, but we know that, when an adult's pre-frontal lobes are damaged, it seems to make little difference to his functioning, except that he becomes coarser. In children, on the other hand, pre-frontal damage causes an obvious drop in intelligence: that is, children *use* the pre-frontal lobes. Could this explain why children experience the "glory and the freshness of a dream", while adults live in an altogether drearier world—that adults have ceased to use this "visionary" function of the pre-frontal lobes?

In *The Philosopher's Stone* I posit a brain operation that is able to restore the "glory and the freshness" to the pre-frontal lobes. Whoever has this operation experiences a kind of revelation. The world becomes alive and exciting and infinitely fascinating, a place of constant "magic".

The underlying assumption here is that the rational intellect—the left brain—is to blame for the dullness of everyday consciousness, with its accompanying sense of triviality and futility. The dullness and rationality are *necessary* if we are to deal with the complexities of adult life; but we somehow *forget* the reality that lies behind our systems of abstraction. And since our vitality is fed by the sense of reality—and purpose—this forgetfulness causes a gradual withering-away of some essential faculty, just as blindness would cause a gradual forgetfulness of the reality of colour. The pre-frontal operation remedies this forgetfulness, generating a sudden enormous sense of the purpose of human existence.

One of the central scenes of the novel occurs when the hero is seated in a Stratford garden, basking in the peace and serenity, and enjoying the sense of timelessness that Elizabeth Fox experienced after a month of meditation on Cephalù. He finds himself wondering idly what this garden would have looked like in the age of

Shakespeare—then suddenly realises that he *knows* the answer; that he possesses a faculty that can tell him exactly what he wants to know. In writing this scene, it struck me as quite obvious that if one could retreat into a deep enough state of serenity, all such questions would become answerable. Yet I was fully aware that "insight" can deal only with questions of a logical nature, not with those involving particularities or facts (e.g., no amount of insight could normally tell me the name of Cleopatra's great grandmother: I have to turn to the history books).

When I thought about this question, it seemed to me that the answer lay in something we know intuitively about states of deep serenity. And this "something" is probably the notion I have already discussed in connection with Buchanan and psychometry: the feeling that the world contains an infinitude of information, and that we possess, although we seldom use, the senses to make use of it. *If* psychometry works—and there is an impressive body of experimental evidence that it does—it must be because objects somehow record everything that has ever happened to them. But we have already noted that our brains also record everything that has ever happened to us. At this point we should observe that, no matter how much information we have access to, we can make use of it only by cross-checking it with information inside us (e.g., faced with a broken-down car, a man who knew nothing about cars would be helpless, even if he had a massive handbook on cars; before he can make use of it, he needs to have certain basic information about cars *inside* his brain). But with an infinitude of information outside us, and something like an infinitude inside us, we possess the basic necessities for answering almost any question.

I am still by no means certain that this "paradigm" is the answer. How, for example, can it explain something that happened to a musician friend of mine, Mark Bredin, as he was travelling back late one night by taxi along the Bayswater Road? Suddenly, he felt certain that, at the next traffic light—Queensway—a taxi would jump the lights and hit them, side-on. But it seemed absurd to tap the driver on the shoulder and say "Excuse me, but . . ." So he said nothing. At the next traffic light, a taxi ignored a red light, and hit them sideways-on . . . Could it have been some kind of extrasensory perception that told him of the approach of the taxi along Queensway at a certain speed, and that the impatient driver would arrive just as the light was turning red?

All that *does* seem clear is that Bredin was tired and very relaxed but that, after a concert, his senses were still alert. The great roaring machine of everyday awareness, with all its irrelevant information, had been switched off and he could become aware of normally-unperceived items of knowledge.

A Ladder of Selves

It was after writing *The Occult*, and while I was working on my book *Mysteries*, that I became aware that the problem was probably complicated by another factor. My discovery that I could use a dowsing rod, and that it reacted powerfully in the area of ancient standing stones, made me clearly aware of this "other" me, the non-ego, who lives in the right hemisphere. I also became increasingly interested in the work of that remarkable man, the late Tom Lethbridge, a retired Cambridge don who studied the use of the pendulum in dowsing for various materials. After exhaustive experiments, Lethbridge concluded that the pendulum responds, *at various lengths,* to every known substance in our world; i.e., that in the hands of a good dowser a 14-inch pendulum will go into strong gyrations over sand, while a 25-inch pendulum will detect aluminium. But, having established this to his own satisfaction, Lethbridge was astonished to discover that the pendulum would respond equally definitely to feelings and ideas; i.e., that a 10-inch pendulum would respond to the thought of light or youth, while a 29-inch pendulum would respond to danger or yellow. This seemed to connect with another baffling phenomenon, which I myself have witnessed: map dowsing. It sounds preposterous, but some dowsers are able to locate whatever they are looking for over a map as well as over the actual area of ground. "Professor" Joad, a confirmed sceptic, described in a Brains Trust programme how he had seen a map dowser accurately trace all the streams on a map from which they had been removed. I have seen a map dowser, Bill Lewis, accurately trace the course of an underground waterpipe on a sketch map drawn by my wife.

And at this point I became fascinated by another equally strange phenomenon, that of "multiple personality". There are dozens of recorded cases of patients who slip in and out of a series of totally different personalities. One of the most widely publicized was described in the book *The Three Faces of Eve*. In *Mysteries* I have described in detail the equally strange cases of Christine Beauchamp and Doris Fischer. In her book *Sybil* Flora Schreiber has described the case of a girl who had sixteen different personalities. Such cases actually look like old-fashioned accounts of "demonic possession". The resident personality, so to speak, is suddenly expelled from the body, and a stranger takes over. When the "resident personality" comes back, he (or she) has no memory of what has taken place in the meantime.

What interested me about such cases is that the various personalities seem to have a definite pecking order or hierarchy, with the most powerful at the top, the next most powerful next to the top, and so on. (The "resident personality" is usually about halfway down the ladder.) Moreover, the "top" personality knows all about all those underneath; the next one down knows about all those underneath, but *not* about the one above. And so it goes on, with the bottom-most personality knowing only about himself/herself.

I made another intersting observation. In many cases, the "top" personality is a more mature and balanced individual *than the patient has ever had the opportunity to become.* For example, Jung's cousin, who was such a case, was a teenager; yet her "top" personality was a mature woman at least ten years older.

In 1973, my own experience of "panic attacks", brought on by overwork and stress, suggested a further insight: that we are basically *all* multiple personalities, although, in well balanced human beings, the others never actually unseat the resident personality. In my panic attacks, I found that I could gain a measure of control by calling upon what seemed to be a higher level of my own being, a kind of "higher me". This led me to wonder how many "higher me's" there are. *And* whether the solution of some of these mysteries of paranormal powers—like precognition—may not lie in this higher level of "myself". In short, whether, as Aldous Huxley once suggested, the mind possesses a superconscious attic as well as a subconscious basement—a superconscious mind of which we are unaware, as we are unaware of the subconscious. My own picture of the "ladder of selves" seemed to suggest that the attic has several storeys.

Lethbridge had begun to formulate a similar theory to explain the accuracy of his pendulum: that there is a part of the mind that knows the answer to these questions, but which can communicate only indirectly. This, of course, sounds more like the right cerebral hemisphere than the "superconscious mind". But then, the right cerebral hemisphere might well be the "seat" of the superconscious mind, if such a thing exists.

Of course, we are all aware that we develop into a series of different people over the course of a lifetime. But we say this is "only a manner of speaking". Is it, though? Some people experience a total personality change when they get behind the wheel of a car; they feel as if a more reckless and impatient "self" had taken over their body. A person involved in lovemaking for the first time may find that he/she is "taken over" by another self, with its own bioligical purposes, and that he/she suddenly becomes oddly self-confident and purposeful. A mother holding her first baby is startled to feel a kind of archetypal mother inside herself taking over her responses and her mind . . .

This leads me to speculate that we may all begin life as a whole series of selves, encapsulated like those Japanese paper flowers, waiting for the right moment to unfold. Someone who never loses their virginity, a woman who never becomes a mother, never allows that particular self to enter the world of the living. Yet a priest who becomes a saint may allow still higher "selves" to unfold, while the rest of us remain trapped in a routine of getting and spending. A Queen Elizabeth or Florence Nightingale may develop areas of her being which remain unconscious in the satis-fied housewife.

All this seems to provide a possible explanation for Alan Vaughan's experience, when "Z" drove the Nantucket "spirit" out of his head. He obviously felt an immense and boundless relief, an explosion of sheer delight. Could this have lifted him, as it were, to a higher rung on the "ladder of selves"? For one thing is perfectly clear: the "lower" we feel, the more we are subject to time. At the beginning of a railway journey, I may feel so concentrated and absorbed that I can simply look out of the window, and experience all kinds of interesting insights and sensations. Later on, I feel less absorbed, but can nevertheless find pleasure in a book. If the journey is far too long, and the train breaks down, and I get cold and hungry, all my concentration vanishes, and time now drags itself slowly, "like a wounded snake". The less absorbed I become, the slower time passes. It seems, therefore, reasonable to assume that if I could reach some entirely new level of delight and concen-tration, time would virtually disappear. In such a state, I might well know what was passing in other people's minds, and know the future. At all events, it seems clear that psychological time is closely related to our control over our own inner states. It seems likely that someone who had achieved a perfect level of collabor-ation between the right and left hemispheres, instead of the present mutual misunderstanding and confusion, would be able to slow time down or speed it up at will. Therefore, whatever we know or do not know about time, one thing seems certain: that increased understanding of our own latent powers will bring increased insight into the nature of time. We shall discover that Wells' Time Machine is the human mind itself.

Coda

What general conclusions does all this enable us to reach?

St Augustine said about time: "When I do not ask the question, I know the answer"; a comment whose meaning becomes crystal clear in the light of what we know of the right and left sides of the brain. The nature of time can be grasped by intuition, but it eludes thought. The author of *The Cloud of Unknowing* was making the same kind of point when he said: "By love may He be gotten and holden; by thought never."

But if categorized knowledge allows us to say little about the nature of time—or of space, for that matter—it at least allows us to glimpse the answer to some of the problems that have preoccupied modern philosophers from Kierkegaard to Sartre and Heidegger. And most existentialist thinkers would regard the problem of time as a part of the question of Being itself. For existentialism, the central problem is the problem of absurdity, the apparent meaninglessness of human existence. Who am I? What am I doing here? Kierkegaard asked: "What is this thing called the world? . . . Who is it that has lured me into this thing, and now leaves me there? . . . How did I come into the world? . . . Why was I not consulted. . . ? If I am compelled to take part in it, where is the director? I would like to see him." The most extreme form of this sense of "absurdity" is Sartre's "nausea", a sense of being negated by the crude reality of objects. But I have already pointed out that this is essentially a left-brain reaction to the world. The same kind of thing happens if you begin to think about some action that you normally carry out instinctively—you can put a good darts player off his game by saying: "How do you hold a dart when you throw it?" The left detaches itself, so to speak, and puts the right off its stroke. The world as seen by the left brain becomes "alienated". The alienation is caused by its detachment from the right, with its intuitive processes. But the left is unaware that it has become detached. It regards itself as the "I", the ego, and cannot imagine that it is somehow incomplete without the backing of another invisible "I". Yet, because of this misunderstanding, it finds itself alone in an absurd Universe, dismayed by the question: "Who am I?" The correct answer is: an incomplete self, the partial mind.

This brings another interesting insight: that our intuitive life

Marcel Proust (1871–1922), the French writer whose great work *A la recherche du temps perdu* displayed a unique view of time past.

seems to be grounded in a sense of security and value. William Kimmel spoke of the problem of modern Man as being "alienation of beings from the source of power, meaning and purpose". This alienation, we have seen, is basically a left-brain misunderstanding. And the "source of power, meaning and purpose" seems to be somehow associated with the intuitive consciousness of the right brain. The playwright Harley Granville Barker called it "the secret life", and pointed out that it is the wellspring of purpose in all men.

In short, it would seem that one of our basic difficulties is our rational, left-brain approach to our fundamental problems—including the problem that is the subject of this volume, time. Does this mean that all attempts to think about such problems are doomed to failure? Fortunately not, since thinking involves intuition as well as rational analysis. The problem is simply to recognize the importance of the rôle of intuition, and to guard against impeding its development by clumsy rationalization. It is easy to see, for example, that Proust might have spent days *thinking* about his childhood in Combray without that sudden intensity of insight brought by the madeleine dipped in tea. The incident of the madeleine revealed that another approach was required—an approach which led to the writing of *A la recherche du temps perdu*. Proust's insight led him to abandon the discursive method, which was the basis of his early work, and to make an attempt at the direct development of "Faculty X". The fact that he was unsuccessful is beside the point. What matters is that he selected an approach that allowed him to say something interesting and valid about the nature of time.

We would do well to ponder his example.

CW

About the Authors

R.S. Porter MA PhD is Director of Studies in History and Philosophy of Science at Christ's College, Cambridge, Director of Studies in History at Churchill College, Cambridge, and Dean of Churchill College, Cambridge. Aside from numerous papers and articles, he has published *The Making of Geology: Earth Science in Britain 1660—1815* (1977) and edited *William Hobbs's 'The Earth Generated and Anatomized' (1715)* (1979) and, with Dr L.J. Jordanova, *Images of the Earth: New Essays in the History of the Geological Sciences* (1979). He is currently writing *The Social History of England in the Eighteenth Century* (for publication 1980).

Richard Knox CEng MIEE FRAS, after qualifying as a Chartered Electrical Engineer, returned to his earlier ambition of being a science writer. For several years he was the Features Editor of *Electrical Review*, the leading British weekly for the electrical industry, and more recently he has taken up the post of Editor of *Nuclear Engineering International*. As well as countless articles on engineering science, he has written about his other lifelong interest, astronomy, in many publications: his three books to date on the subject are *Experiments in Astronomy for Amateurs* (1975), *Discover the Sky with Telescope and Camera* (1976) and *Foundations of Astronomy: From Big Bang to Black Holes* (1979).

Chris Morgan BSc took his degree in economics at the University of London, after which he joined one of Britain's largest industrial companies. After rising to the position of Purchasing Executive he decided to resign and concentrate full-time on writing. For some two decades he has been amassing a vast library of science fiction, science "fact" and speculation from the past; he has published a number of science-fiction stories. His first book, *Future Man: The Further Evolution of the Human Race* is to be published in 1980.

E.W.J. Phipps BSc took his degree in biochemistry at Exeter College, Oxford, and is currently employed as a writer for a British pharmaceutical company. His special interests are in the general areas of biochemistry, animal psychology and small-systems computing. He is currently at work on his first book, which is closely linked to these special interests, *Immortality: The Search Continues*.

Iain Nicolson BSc FRAS FBIS is a Lecturer in Astronomy at the Hatfield Polytechnic. He is well known as a writer on astronomy and space science: his books include *Astronomy: A Dictionary of Space and the Universe* (1977), the exceptionally well received *The Road to the Stars* (1978) and, with Patrick Moore, *Black Holes in Space* (1974). He is currently at work on a number of projects including *Gravity, Relativity and the Black Hole*, to be published in 1980.

Brian John MA DPhil was a student of Jesus College, Oxford; while there he was joint leader of the University expeditions to Iceland (1960) and East Greenland (1962). With the British Antarctic Survey he spent the summer of 1965–66 in Antarctica on research concerned with glaciation and land and sea-level changes in the South Shetland Islands. On his return to the UK he joined the geography department of Durham University. He resigned in 1977 to concentrate on writing and to found his own small publishing company. For *other* companies he has written *Pembrokeshire* (1976), *Glaciers and Landscape* (with David Sugden; 1976), *The Ice Age: Past and Present* (1977), *The World of Ice* (1978) and edited *The Winters of the World*.

Colin Wilson became internationally celebrated overnight with the publication of his first book, *The Outsider*, in 1956. His many novels include *Ritual in the Dark* (1960), *Adrift in Soho* (1961), *The World of Violence* (1963), *Man Without a Shadow* (1963), *The Glass Cage* (1967), *The Mind Parasites* (1967), *The Philosopher's Stone* (1971), *The God of the Labyrinth* (1970), *The Schoolgirl Murder Case* (1974) and *The Space Vampires* (1975). His plays are *Viennese Interlude* (1960), *Strindberg* (1970) and *Mysteries* (1979). Apart from *The Outsider*, his works of non-fiction include *Encyclopedia of Murder* (with Pat Pitman; 1961), *Origins of the Sexual Impulse* (1963), *Beyond the Outsider: The Philosophy of the Future* (1965), *Sex and the Intelligent Teenager* (1966), *Introduction to the New Existentialism* (1967), *A Casebook of Murder* (1969), *Poetry and Mysticism* (1970), *The Occult* (1971), *Order of Assassins* (1972), and *Mysteries* (1978).

Acknowledgements

We would like to thank the following for kindly permitting the use of copyright material:

J. Allan Cash, 62; Heather Angel, 130 (top and bottom), 137 bottom; Australian National Radio Astronomy Observatory, 193 (bottom); BBC Hulton Picture Library, 16, 19, 23 (top), 37, 41 (top), 42, 47, 50, 52, 57, 155 (bottom), 159, 170, 244, 258, 287, 299 (top), 314; *from* De La Beche, H.T., *Sections and Views Illustrative of Geological Phaenomena*, (London, 1830), 31; Bibliothèque Nationale, Paris, 14; Bodleian Library, Oxford, 106 (right); Trustees of the British Museum, 9 (photo John Freeman), 11, 23 (bottom), 276 (BM, Natural History), 282 (BM, Natural History); Prof R. Cooke, 252; G.R. Coope, 273; Crown Copyright, Central Office of Information, 125; Crown Copyright, National Physical Laboratory, Middlesex, 126 (bottom); Crown Copyright, Science Museum, London, 93, 94, 97 (whole page), 98, 99 (whole page), 101, 105, 106 (left), 108, 111 (top left), 117 (bottom), 122 (right), 126 (top); CSIRO, 65 (bottom); Dept of Mines, Energy and Resources, Canada, 243; Deutsches Museum, Munich, 86; Don L. Eicher, *Geologic Time* (Prentice-Hall Institute), (p. 119), 267, (p. 128) 269; Mary Evans, 72; Mary Evans/Society for Psychical Research, 286 (left and right); painting by Andrew Farmer, 234; Follett, B.K. and Sharp, P.J., *Nature* (London, 1969), (**223**, p. 968), 143; E. Fromm, in Olsson, I. (ed) *Radiocarbon Variations and Absolute Chronology* (Nobel Symposium 12, Stockholm, 1970), 275; German Royal Academy of Sciences, 87; Photo Giraudon/The Louvre, Paris; SPADEM, 41 (bottom); Photo Giraudon, The Louvre, Paris, 85; photographs from the Hale Observatories, 158, 184, 189, 213 (top); Helwan Observatory, Egypt, 67; by courtesy of Mrs Arthur Holmes, 259; Institute of Geological Sciences, 249; Brian John, 278; Kitt Peak National Observatory, Arizona, 81 (top); Laboratory of Tree-ring Research, University of Arizona, 279; Lick Observatory, California, 76 (whole page), 81 (bottom), 213 (bottom); B. Lofts and A.J. Marshall (Ibis, 1960), (**102**, p. 209), 133; Brian Loomes, 114, 119, 121; The Mansell Collection, 7, 26–7, 43; MGM Pictures, 289; Museum of History of Science, Oxford, 102; Courtesy of the Trustees, The National Gallery, London, 25; Picturepoint, London, 6; Ann Ronan Picture Library, 46 (bottom), 48, 49, 51, 56 (top and bottom), 59, 70 (top), 193 (top), 240; Royal Greenwich Observatory, 46 (top), 65 (top); Royal Observatory, Greenwich, 70 (bottom); RSPB/Fritz Polking, 134; Science Museum, London, 103, 112 (top and bottom), 122 (left), 124; lent to the Science Museum, London, by the late P.L. Harrison, 111 (right), 117 (top); lent to the Science Museum by W.H. Barton, 115; Studio Jon, 280; The Tate Gallery, London, 239; John Topham Picture Library, 146 (top and bottom), 147, 155; UKAEA Harwell, supplied by R.L. Otlet, Carbon Dating 14 Laboratory, 272; University of Exeter, 271; US Geological Survey, 255.

Index

absolute time, 36, 39, 40, 54, 157–62, 165, 171, 173, 180, 195, 214, 233, 234
Agassiz, L., 242
ageing, 152–6, 180
animals' time sense, 135–6
Aristotle, 12, 21, 28, 35, 163, 237
Arnold, J., 118
Arrhenius, S., 145
Aschoff, J., 141
astrology, 58, 59, 71–4, 79, 82
Augustine of Hippo, St, 15, 313
Avicenna, 237

Bacon, F., 21, 22, 24
Bacon, R., 109
Bailey, 183
Bain, A., 122, 123
Barrow, I., 35, 36, 160
Becquerel, H., 31, 258
bees' time sense, 132, 134–5, 139, 148
Bennett, J.B., 292–3
Bergson, H., 39, 43
Berkeley, Bishop, 39
Beschel, R.E., 278
Bessel, 82
biorhythms, 129–45; human, 140–5
birds' time sense, 132–3, 139
Birrell, N.D., 221
black holes, 214–21, 227, 234
Boltwood and Holmes, 245, 258–9, 261
Boltzmann, L., 231
Braginskii, 203
Brahe, T., 55–6, 57, 58
Bredin, M., 309
Breguet, L., 120–1
Buchanan, J.R., 286, 293, 294, 309
Buckland, W., 238, 240, 242
Buffon, Comte de, 30, 32, 245
Bünning, E., 131–2, 142
Burnet, T., 28

Cailleux, Prof, 253
calendar,
    early, 10, 84–94, 159–60
    Gregorian, 10, 90, 110
    Julian, 90, 93, 109
Carnac, 90
Catherine the Gt, Empress, 98, 118, 120
Charpentier, Jean de, 242
Chesterfield, Lord, 38
Christenson, J., 232
Christianity, time in, 13–22, 24, 160, 236, 237–8
Clavius, C., 110
Clay and Crouch, 190
Clement, W., 113
clock,
    atomic, 71, 126–8, 161, 183, 211

development of, 36–7, 94–128
    electric, 122–5
    famous, 104, 106
    grandfather, 113
    incense-, 99
    mechanical, 36, 96, 102–23
    quartz, 123, 125
    sand-, 101
    shadow-, 94–5, 98
    sidereal, 52
    sundial, 50, 58, 59, 63, 64, 67, 95–8
    water-, 95, 98–9, 102, 104
clock paradox, 183–9
clock time, 7, 36, 40, 52, 64
Colburn, Z., 301, 304, 305
Copernicus, 21, 28, 48, 50–1, 54, 55, 58, 68, 222
Crowley, A., 303
Cuvier, G., 33, 238

Dante, 102
Darwin, C., 30, 33, 34, 35, 244, 256, 258
Darwin, E., 33
dating the past,
    early beliefs, 236–9
    Earth's age, 255–7
    geological time, 239–48
    landscape study, 248–55
    modern methods, 261–84
Davies, P.C.W., 221, 229, 233
Davis, W.M., 248, 250–1, 252
day, 10, 27, 45, 50, 52, 54, 77, 84
De Candolle, 131
De Coursey, P., 135
De Geer, G., 274, 275
De Mairan, 130
Denton, W., 286, 293, 294
Descartes, R., 28, 30
Dicke, 203
Divi, S., 151
Dixon, J., 299
Dondi, 104, 106
Donne, J., 5, 15, 83
Duhamel, 131
Dunne, J.W., 40, 287, 291, 294
Duras, M., 43
duration, 36, 39, 148–50, 158

Earth,
    age of, 28–34, 255–71
    distance from stars, 80, 82
    rotation, 68–9, 71–3, 79, 165, 226
    Sun and, 45, 47, 48, 50, 51–6, 58–64, 68, 82, 165, 226
Eckhardt, Meister, 18
Eddington, Sir A.S., 207
Ehret, C., 138
Eicher, D.L., 260, 283
Einstein, A., 40, 58, 82, 162, 169–70, 178, 196, 201–2, 205, 207–9, 212, 219–20
Eliot, G., 39
Ephemeris Time, 75, 77, 80

Esmark, J., 242
Essen, L., 128
eternity, 12, 16–17, 236–7

Fabergé, C., 123
Fairley, P., 290–2, 304–6
Fitzgerald, G., 169
Fliess, W., 144
Fontenelle, 29
fourth dimension, 194–200, 233
Freud, S., 39, 144, 150
Frisch, K. von, 134–5
Fritts, H.C., 277
future, 5, 6, 25–6, 152, 291
  absolute, 198–9
  Christianity and, 22, 24
  knowledge of, see precognition
  past and, see under past

Galileo, 21, 28, 30, 36, 52, 110, 111, 202, 212
Garner and Allard, 131
Gasset, Ortega y, 25
Geikie, A., 242
George III, 116, 118
Gissing, G., 307–8
Godly, J., 290, 291
Goldsmith, O., 113
Graham, G., 114, 115
Graves, R., 300–1
Greenwich Mean Time, 66–8, 75, 123
Gregory I, Pope, 95
Gregory XIII, Pope, 109
Grimthorpe, Lord, 121
Guarnier, Canon, 307

Hafele and Keating, 183
Halley, E., 30, 67
Harker, J., 136, 138
Harrison, J., 115–16
Hasan, A., 96
Heim, A., 150
Helmholtz, H. von, 245
Henlein, P., 107
Herrick, 17
Hertz, H., 166
Hilbert, 296
Homer, 18, 21
Hooke, R., 113, 244
Hope-Jones and Boswell, 123
hours, 8, 92–4
Hubble, E.E., 222
Hume, 39
Hurkos, P., 295, 298
Husserl, E., 307
Hutton, J., 32, 239–44, 248, 251, 256–8
Huxley, A., 25, 311
Huygens, C., 36, 113, 116

James, W., 293
Jansky, K., 192
Joachim of Fiore, 16

Joad, C.E.M., 285, 310
Johnson, Dr, 22, 296
Joyce, J., 42, 43
Judaism, time in, 13, 15, 236

Kant, 29, 30, 39, 306
Kelvin, Lord, 31, 245, 258
Kennedy, President, 299
Kepler, 21, 28, 55, 57–9, 68, 80, 92
Kerr, R.P., 220
Kimmel, W., 314
King, L., 252
Kramer, G., 132–3
Kruscal, M.D., 220

Lamarck, 33
Laplace, P. de, 29, 214, 215
Layzer, D., 232
Leakey, L., 34
Leibniz, 30, 162
Leonardo da Vinci, 107, 237
Lethbridge, T., 310–11
Libby, W.F., 272
Linnaeus, 33
Locke, J., 36, 38
Lorentz transformation, 169, 175–80, 196
Luther, 17
Lyell, C., 32, 242–4, 248, 251, 256–8

McCarthy, 156
MacLuhan, M., 38
Marrison, W.A., 128
Marx, K., 22
Maxwell, J.C., 165–6
mean time, 64, 66–8, 75, 123
memory, 5, 39, 42, 151–2, 293, 295
Meton, 88
Michelson-Morley experiment, 166–9, 170
Milne, E.A., 230
Minkowski, H., 42, 194–6, 233
minute, 36
Moberly and Jourdain, 285–7, 297, 308
month, 10, 53–5, 77, 79, 84, 86
Moon, 10, 26
  biorhythms and, 144
  Earth's rotation and, 68–9, 71, 77
  eclipse, 78, 79, 86
  motion, 45, 47, 53–6, 75, 76, 78
  phases, 76–8, 79
More, Sir T., 23–4
Mozart, 151
Mumford, L., 36
Murchison, 248

Newton, 21, 28, 35, 57, 58, 67, 69, 157–65, 171, 194, 201–3, 212, 233, 234
Nostradamus, 299

Ogden, T., 119
O'Neill, J., 296–8, 308
Orwell, G., 25

past, 5, 6, 152, 234, 291
  absolute, 198–9
  chronology of, 18–20
  dating, *see* dating the past
  Earth's timescale, 236–84
  future and, 160, 161, 192, 194, 230, 287
  present and, 21–2, 230
  psychic experience of, 285–7, 293, 295,
    297–8, 307–9
Penck, W., 252
Penfield, W., 293, 295
Penzias and Wilson, 226
Pepys, S., 36, 37
Petrarch, 21
Piaget, J., 6, 148
Plato, 12, 13, 21, 39, 83, 92
Playfair, J., 32, 242
Poincaré, H., 169
precognition, 40, 287–93, 295, 298–9,
  304–7, 309, 311
present, past v., 21–2, 230
Priestley, J., 24
Proust, M., 43, 305, 314
Ptolemy, 21, 47–8, 50, 51, 54
Pythagoras, 92

relativity, 6, 40, 82, 161, 162
  general, 157, 184, 200–12, 233
  special, 157, 169–78, 190, 200, 206, 212,
    225, 233
Rhine, J.B., 306
Riemann, G.F.B., 208
Rittenhouse, D., 120
Robbe-Grillet, A., 43
Roemer, O., 166
Rossi and Hall, 182–3
Ruskin, J., 30–1
Rutherford, E., 31

Saros cycle, 79, 159
Sartre, J.-P., 296, 313
Schreiber, F., 310
Schwarzchild, K., 214–16
Schweiger, H., 139
Scrope, G.P., 32
second, 36, 77, 113, 128
Sedgwick, 248
Seneca, 12, 237
Shakespeare, 8, 17, 21, 39, 309
Shortt, W.H., 122, 123
sidereal time, 52, 66, 68, 75
simultaneity, 6, 40, 161, 165, 171–4
Sixtus IV, Pope, 109
Smilley, F.F., 301
Smith, W., 241–2, 248
Soal-Shackleton experiments, 306
space,
  absolute, 160–2, 165–6, 171, 180, 195,
    227
  infinite, 28
  time and, *see under* time
Sperry, R.W., 301

Steno, N., 239, 248
Sterne, L., 38
Stonehenge, 5, 6, 90–1
Strehler, 154
Strughold, H., 142
Sturt, 151
Sun, 10, 11, 12, 26, 36
  eclipse, 46, 78–9
  motion, 45, 48, 50–64, 71–5, 77–8, 82
  time by, 54–64
sunspots, 65

Thomsen, C., 34
Thorarinsson, S., 274
time,
  consciousness of, 44, 145–52, 312–13
  direction of, 34, 189, 194, 229–36
  gravity and, 210–14, 233
  mind and, 291–300, 303, 312–13
  motion and, 35–6
  nature of, 10, 12, 82–3, 140, 285, 291–4,
    304, 313
  philosophers and, 12–13
  space and, 28, 91, 161–2, 174, 227–9
  subjective, 36, 39–44
  universe and, 222–9
time dilation, 175, 179–89, 192, 210–11
time-keeping, 36–7
time travel, 192–4, 233–4, 288–94, 306
time zones, 66–7
Tompion, T., 112, 113
Toynbee, A., 304
Tricart, 253

Universal Time, 67–8, 75, 77, 80
Ussher, Abp J., 245, 256, 282

Vaughan, A., 298, 300, 312
Velikovsky, I., 254, 284
Venetz-Sitten, I., 242
Vessot and Levine, 211, 214
Vyvyan, J., 307–8

watches, 108–9, 112–13, 116, 118, 121,
  124, 128
week, 10, 92–3
Wegener, A., 245
Wells, H.G., 42, 93, 288–92, 294, 307
Werner, A, 238
Whitrow, G.J., 160
Willard, S., 120
Wooldridge, S.W., 251
Woolf, V., 41, 42–3
Wordsworth, W., 308

year, 10, 47, 77, 84–7
  Great, 12
  leap, 110
  Long, 91–2

Zeno's paradox, 303
Zodiac, 59, 71–4